The ABC Book of Hi-Fi/Audio Projects

By George deLucenay Leon

TAB BOOKS
Blue Ridge Summit, Pa. 17214

FIRST EDITION

FIRST PRINTING—SEPTEMBER 1977

Printed in the United States
of America

Library of Congress Cataloging in Publication Data

Leon, George, 1914-
The ABC book of hi-fi/audio projects.

Includes index.
1. High-fidelity sound systems--Amateurs' manuals.
2. Sound--Recording and reproducing--Amateurs' manuals.
I. Title.
TK9968.L46 621.389'3 77-22702
ISBN 0-8306-7921-9
ISBN 0-8306-6921-3 pbk.

DEDICATED TO

Sylvia, who did so much of the work and had none of the fun.

ACKNOWLEDGEMENT

Acknowledgement is due to the many people who helped with suggestions and advice. Special thanks are owed to the following manufacturers who were kind enough to allow me to use their printed circuit patterns and layouts which appear in the book:

GC Electronics of Rockford. Illinois. for their liberal cooperation in making parts available for experimenting.

Motorola Semiconductor of Phoenix. Arizona. for the circuit pattern which appears in their application note AN-484-A.

National Semiconductor of Santa Clara. California. for the generous help they gave me with circuits for the LM 381.

Signetics Corporation of Sunnyvale. California. for allowing me to use the pattern and layout for the 35-watt amplifier as extracted from their Linear Data Book.

FOREWORD

There were two goals in presenting "The ABC Book of Hi Fi Audio Projects." First, to assemble a collection of useful circuits to meet a variety of needs and projects that would be of interest to a broad range of hobbyists. Obviously, to be of interest, the projects had to be of varying degrees of complexity and would have to be flexible enough to lend themselves to added experimentation and optimization. Therefore, they were brought together with the thought that several could be assembled together to make a more complex and bigger project—one in which the whole was greater than the sum of its parts. For example, the microphone mixer is such a project and consists of several smaller projects being modified to suit the overall unit. Another example, by using the formulas in Chapter One to design a power supply, then adding filters and attenuators, an excellent unit can be built that the reader can really be proud of. And he will have designed it!

One more example: Even if you don't need a mixer, that same circuit can become part of a low-distortion, quality, stereo preamp using the latest in IC technology. By adding one of the amplifiers shown in Chapter Five, you have a fine sound system to suit your particular needs.

This "bite-size" approach to building allows the not-so-very-experienced hobbyist to build large and somewhat difficult projects without developing "brain fever."

Although ICs are used for several projects, transistors were not forgotten, as they are still very much with us. Commercial users as well as hobbyists continue to depend on the transistor. Some of the circuits, as an example, use FETs to update a tube circuit: these should provide the experimenter with food for thought in still more updating with the use of ICs for replacing older circuits.

This book is an attempt to fill the wide gap that exists between the kit builder (who blindly follows the manufacturer's instructions with even the lengths of wires specified) and the advanced hobbyist who has such a grasp of electronics that he can evolve a circuit from the application notes alone. In the first case, the kit builder has learned almost nothing at all about the theory behind the project he is building, and in the other the appeal is limited to engineers, technicians, or very advanced experimenters.

This is not to deny the value of kit building. Kits fulfill a definite need. The experimenter who wants a sophisticated piece of equipment can build it with a substantial saving over the cost of the assembled unit by doing the labor.

The second goal of this book, a corollary of the first, was to show the beginner, or at least one not entirely certain of himself, how to work out a completed project from a schematic alone.

Professionals use mock-ups, once a breadboarded circuit has been evolved. Why shouldn't the hobbyist use the same technique? In many electronics magazines the size of the chassis for a project in a construction story is given—even the places where the holes must be drilled. But when that same experimenter is faced with only the schematic where does he go from there? With a mock-up mistakes are avoided. If mistakes are made they occur in the mock-up stage, costing practically nothing. This is better then spending hours drilling a metal case which, by a mistake, can be rendered virtually useless.

The use of a mock-up produces a better looking, more useful piece of equipment, with a carefully ordered layout. The mock-up is a definite help to dress leads in a better fashion, eliminating long leads with the consequent oscillation, hum, and the other troubles that tend to haunt the careless builder. By building a copy of the finished unit with materials found in the home, the end product will reflect the care lavished on it and be an object of pride. A better built unit is a better working unit.

The second section of this text is devoted to constructing the most needed test instruments that the hobbyist is wont to use. With the schematics provided, the reader can build his own equipment at a fraction of the cost of commercial units, enjoy the challenge, and learn the theory behind what he is building.

The schematics should be regarded as springboards for the reader to inspire him to experiment further.

So—good building!

George deLucenay Leon

CONTENTS

CHAPTER 1
FROM SCHEMATIC TO FINISHED PRODUCT

So, you have a project you want to build, but all you have is a schematic. The first question that comes to mind is, "How do I figure out how large a chassis I need?" Sometimes, a construction article in one of the magazines will specify the chassis size that the author used in his prototype. There may even be a pictorial showing the parts interconnected. But, what happens when all you have staring you in the eye is the schematic? Or, what happens when you want to design your unit differently from the one shown in the article?

How big should the chassis be? Would 7 by 13 inches be better then 8 by 12? Will a 2-inch high chassis be sufficient or would 3 inches be necessary? You can, of course, pick a size you hope will be right and start drilling wherever you think the external controls should go. That's one way to go, but the hard way.

Have you ever had the experience, after spending an evening drilling all the cutouts, of facing the dilemma that once the potentiometers were in position, the printed circuit board could not go in? You then found that you not only wasted time, but now have a case that is virtually useless for anything else. It has happened to everyone at least once. So do it the way the professional designers do it—make a mock-up first. It may take a little more time at the beginning, but you save much time and money in the long run.

MAKING THE MOCK-UP

The mock-up is a full scale cardboard model of what you believe the finished project will look like. Any cardboard will do, corrugated or not. You can cut up a box or you may find a sheet of board large enough for your purpose. The first step is to gather up all of the components that are going into the project. This gives you an approximate idea of the size and shape of the case you will need. The reason for saying that all of the components must be on hand before beginning the mock-up is that, say, a certain meter is unobtainable at the time you want it. So the substitute's size will be different. If you get a smaller meter, not too much harm is done. But what if you must obtain a larger meter? What does that do to your front panel layout? That's why every component should be on the workbench before you begin.

The front panel is going to be laid out first. Let's say that you have two pots, two toggle switches, a meter, and a neon lamp. Place them on a sheet of paper, face up or face down, it doesn't make any difference. You only want to be able to estimate how much room they will take. You feel the meter should go in the center. Place it on the paper. One toggle switch turns on the power. Lower right hand corner? Let's try it. Put that on the sheet. The neon lamp indicates when the power is on, so logically it should go above or near the switch, since the two will be interconnected. The two pots must have enough room around them so that, once the knobs are on the shafts, there will be enough space for the operator's fingers to fit. The other toggle switch might go to the opposite side of the panel to balance the first. You now have all the controls together on the sheet. How does it look? Can you get them closer, or will it be inconvenient? Don't forget that the diameter of the pot may be larger than its knob, so that must be taken into consideration. How much room do they cover? (See Fig. 1-1).

Let us say that once the components are arranged to your satisfaction, the area measured is 2 3/4 inches high by 6 1/2 inches wide. Looking through a mail order catalog, you see that the nearest size to what you need is 7 inches by 3 inches high. (Reputable mail order houses are listed in Appendix Three, with their addresses.) Good, so far. But what about the depth? That is where most of the trouble occurs.

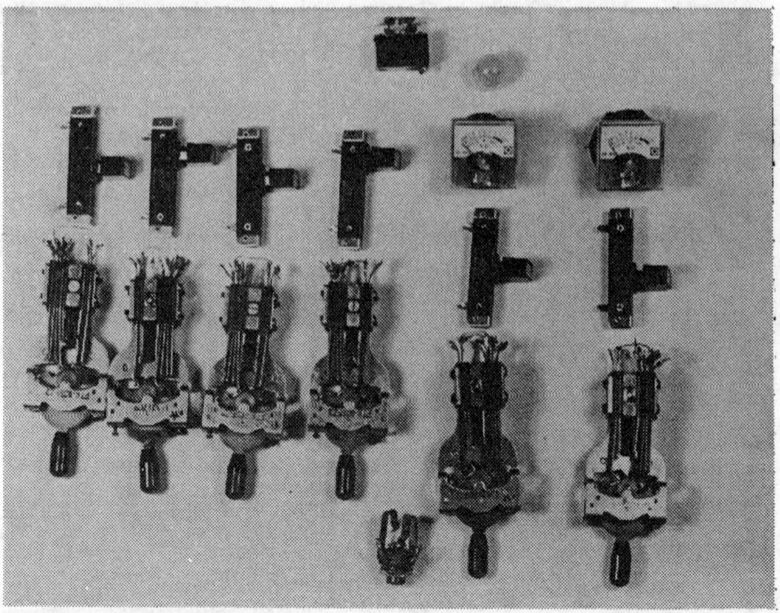

Fig. 1-1. Parts are assembled on a sheet of paper to determine size of chassis necessary for project.

Take another sheet of paper and mark off a 7-inch line. Next, lay out the components that are to fit on the bottom of the chassis. You may have the components already wired to a board or possibly they are breadboarded. This will give you an idea of how much space they will take.

Don't forget that components such as transformers or large electrolytics take up a lot of room and are not easy to maneuver. Also, don't forget that the ON-OFF switch is located on the right hand corner, so your power supply should be as close to it as possible. Don't hesitate to move things around. You might even have to change the front panel. Do it now. It will be too late afterwards.

Remember that you need space to wire a circuit board to its inputs and outputs. You now find that the space covered is about 5 inches deep and 7 inches long. There is a chassis listed in the catalog which measures 5 × 7 × 3 inches. That seems about right, but it is too early to buy it.

Take a sheet of cardboard and mark it off as shown in Fig. 1-2. Cut through on the heavy lines but make only a shallow cut along the dotted lines. Now bend the cardboard along the cuts as in Fig. 1-3.

Tape the corners together using any gummed tape. You now have a case that matches the commercial chassis in every particular except one: a cover. But unless you plan to mount parts on it, it is not necessary.

Using the layout on the sheet of paper that represents the front panel, poke holes in the cardboard for the controls. Fit the parts into the holes—the pots, switches, and lamps. See Fig. 1-4A and 1-4B. On the bottom of the chassis, put in the power supply and the other components. If it is a high gain circuit, are

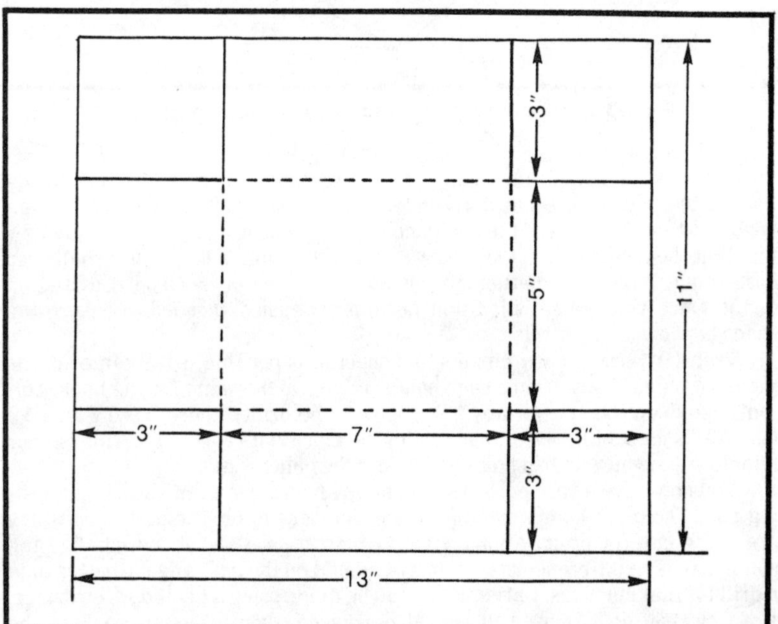

Fig. 1-2. Cardboard marked off for cutting and scoring.

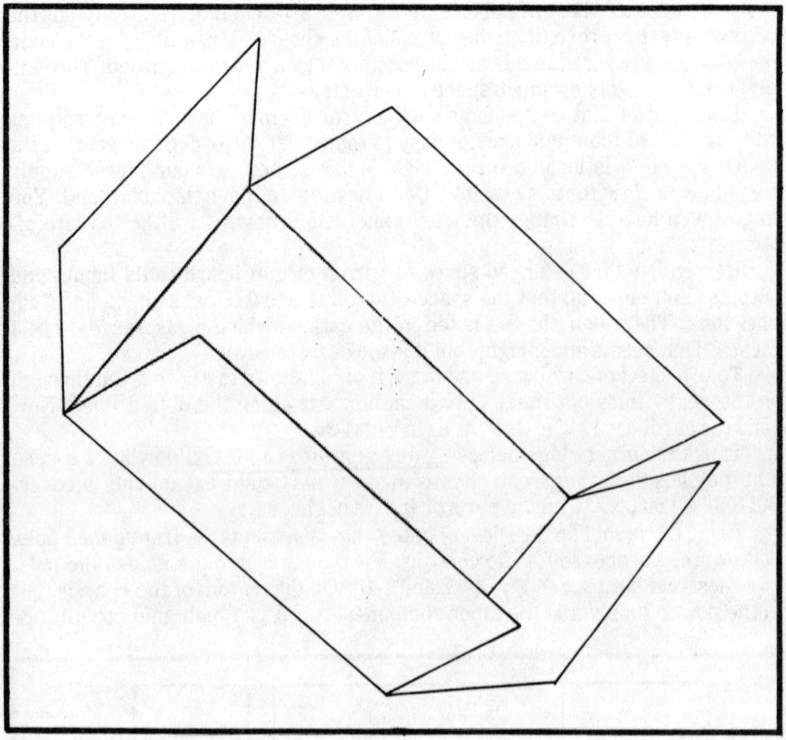

Fig. 1-3. Cardboard cut and ready to be folded and taped.

you sure that the input is far enough from the output so there will be no interaction? Once you have separators under the PC board, will there be enough height so that the leads of the components do not cause a short? You may now find that the neon lamp is too close to a knob. Possibly a longer but shallower case might have been better. So cut another piece of cardboard and start again. Once you are satisfied that no improvement is needed, measure the space between components.

Transfer these measurements to a sheet of paper that is the same size as the front panel. Mark where each hole is to go. Do the same for the back, bottom, and even the sides if any holes have to be drilled there (See Fig. 1-5). Glue each sheet of paper to that part of the chassis it represents. This is your template and you will drill holes just where the paper is marked.

You now have a case perfectly laid out with no unpleasant surprises awaiting you. The more complicated a project happens to be, the more necessary are these careful preparations. Even a plastic case should be handled the same way. The difference is that an awl heated on the gas range is better than a drill for making holes. Larger holes can be made using a heated screwdriver. If an electric drill is used, it should be slowed down to avoid cracking the plastic.

Fig. 1-4A. Parts mounted on cardboard with slits for pots and holes for the meters and other controls.

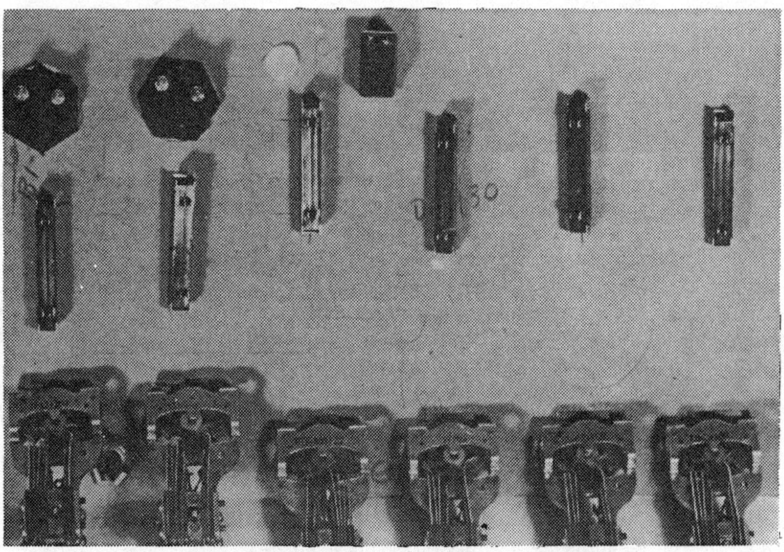

Fig. 1-4B. Rear view of parts mounted on cardboard.

THE FINISHED PRODUCT

Many hobbyists feel that wiring the unit ends the project. They will leave the metal chassis unpainted and sometimes even without indications as to what the controls are supposed to do. That's all right, if you are satisfied with a

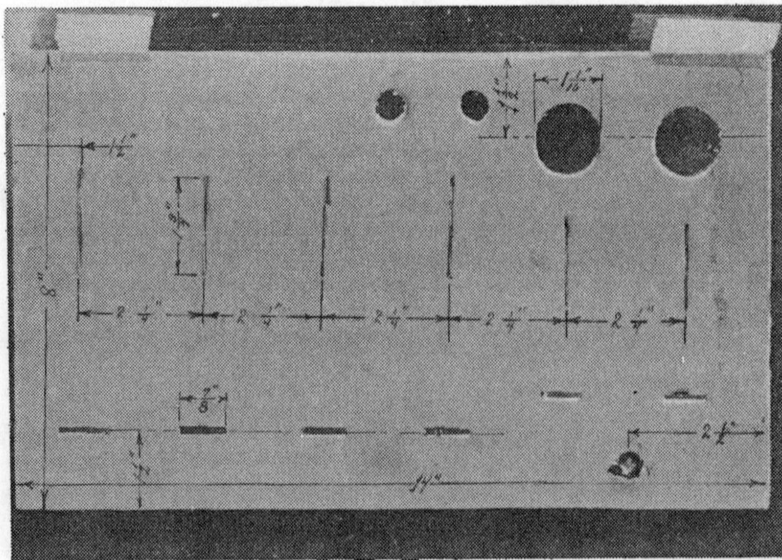

Fig. 1-5. Cardboard cut and measured. It is ready to be used as a template.

home-made piece of equipment. There's a world of difference between hand-made and home-made. There is no reason why a chassis should not be painted attractively. After all, professional units do not come in raw metal. So, do the same for your hand-made equipment.

Once the paint has dried, you should think of labeling the switches and the various controls. There are many options available to the hobbyist. The simplest are the punched-tape labels. There are many colors of tape to match or contrast with the color of the chassis. Engraving tools, sold through electronic mail order houses, allow you to engrave directly into the metal of the chassis. This requires practice to make the engraving come out well. But, there are several transfer types of decals which give the unit a well make look.

Try a carrying handle at the top, or even two on the sides. Look at the professional units and use their design as inspiration. Lots of money and time went into making those units practical and attractive. Why not do the same?

Once the decals, or whatever form of marking, have dried you are ready to wire the unit. Here again, careful workmanship pays off. A cold solder joint can take hours to run down. According to a well-known kit manufacturer, 90% of all kit failures are due to poor soldering. Use a well tinned iron of the proper wattage. For PC boards, no more than 30 watts should be used for soldering. For heavier work, 100 to 140 watts is about right. Make sure the solder flows easily over the connection before removing the iron.

CHAPTER 2

TECHNIQUES FOR MAKING PRINTED CIRCUIT BOARDS

Originally restricted to the manufacturers who used thousands of PC boards. the hobbyists have since discovered their value. With the advent of ICs and high density circuits. the copper covered board became a necessity.

VARIOUS METHODS

There are several ways a hobbyist can make a PC board. One method is to take a fine paint brush dipped in acid resist. which is a type of varnish. and paint on the component connecting lines on a copper coated board and dots where the components' leads are to come through. Also available are pens which are filled with the acid resist and are used like a fountain pen. These are easier to use since they make a much finer line and are easier to control. The lines must be carefully drawn. making certain no lines touch which should not.

Once the resist is dry. the board is placed in a plastic tray containing etching acid. Stirring it once in a while. an etched board should result after 20 or 30 minutes in the acid bath. At the end of that time. all of the uncovered copper is etched away so that only the lines and dots protected by the resist remain.

The board is rinsed in running water so no traces of the acid are left. Be careful that it does not spatter as the acid can burn clothes and skin. Once the board is thoroughly washed. the resist is removed with a thinner. or with nail polish remover. Then the entire board is cleaned with very fine steel wool or kitchen cleanser. What you have left is a faithful reproduction of the layout drawing in copper.

GC Electronics. Lafayette Radio. and Radio Shack all have kits to get you started. The kit contents will include the resist. a pen. some copper coated boards. plus the acid and a resist thinner.

Holes are drilled with bits whose diameters match the component leads. The best way is with a drill press. It can also be done with a hand drill or an electric drill. but certain precautions should be taken. Make certain the board is securely fastened down so there is no possibility of movement.

The next part is easy. Put in the components and solder with a low wattage iron. (See Fig. 2-1). Make sure the diodes and electrolytic capacitors are oriented properly with respect to their polarity. It is also advisable to use some form of heat sinking when soldering semiconductors. They are sensitive to heat. A soldering tip too hot or held too long against a semiconductor lead can ruin it. A thin, fast melting solder is recommended.

METALLIC TAPE

The method described above is useful only if it is a simple circuit and does not contain any ICs...the closeness of their pins make it almost impossible to paint them in properly. There is a company which makes metallic press-down strips and patterns which match the company's uncoated perforated board. (See Fig. 2-2) These metallic pieces are pressure sensitive and are soldered together as one would a wire. Again, a low wattage iron must be used and heavy components cannot be employed with this method, as they pull the tabs away from the board. This method can be combined with point-to-point wiring.

The methods outlined above are only for the times you are making a one-of-a-kind board.

One of the projects in the book. a microphone mixer. requires four boards. Though not exactly alike. the same basic pattern can be used. There is a difference in component count. but basically they are very close. In this case. it would be recommended to use the photographic process which allows for easy duplication.

PHOTOGRAPHIC PROCESS

Basically this process is especially valuable when several boards must be made alike. It consists of making a negative of the pattern by placing the various pattern components on a sheet of film. Several companies make these graphic aids. which are made of black pressure sensitive tape and with different patterns. These are fastened down and the film is used as a negative. The copper clad board is sensitized by painting it with a solution which reacts to light. The film is clamped onto the board in a frame. This is then exposed to a 100-watt bulb. Floodlamps are often recommended. but the smaller lamps will do the same job and are safer to use. It merely takes a little longer exposure to the light. The board is removed from the film and placed in a developing solution which brings out the pattern. The same film negative can be used countless times. Each board is made very quickly once the negative is on hand.

In the methods mentioned above. it was taken for granted that you had a pattern to copy. What if you don't have one?

There is still hope. A schematic lays out the components in a very logical way. This is done to make the function of the circuit easier to understand. But. you can make a PC board by following the schematic fairly closely. Notice the two-transistor schematic in Fig. 2-3. (No values are given since it is only a theoretical example.) Now compare it to Fig. 2-4 which is a sketch for PC board. Figure 2-5 is the resultant board, ready for etch. Yu will see at once the resemblance between the two. Lines in a PC pattern do not have to be straight, but can curve to go around another component. In the above example, there

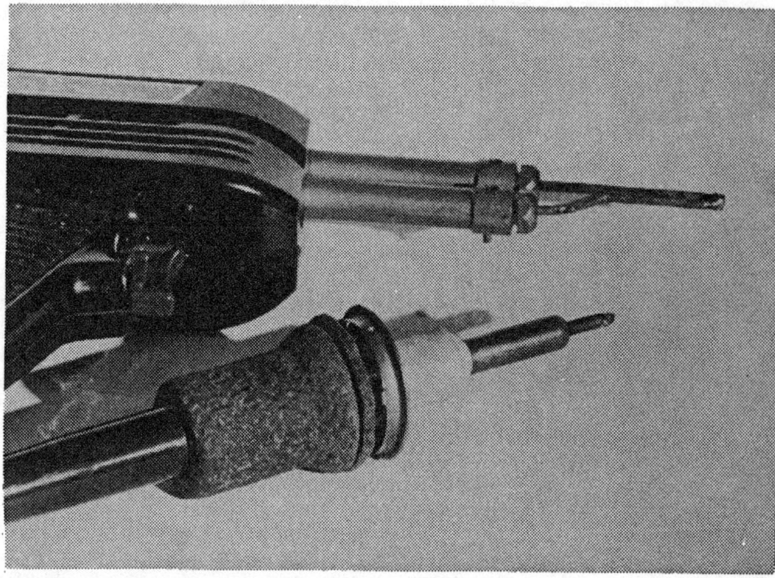

Fig. 2-1. Two types of soldering tools. A dual-heat gun—100 and 140 watts and a small soldering pencil (about 27 watts) for PC work.

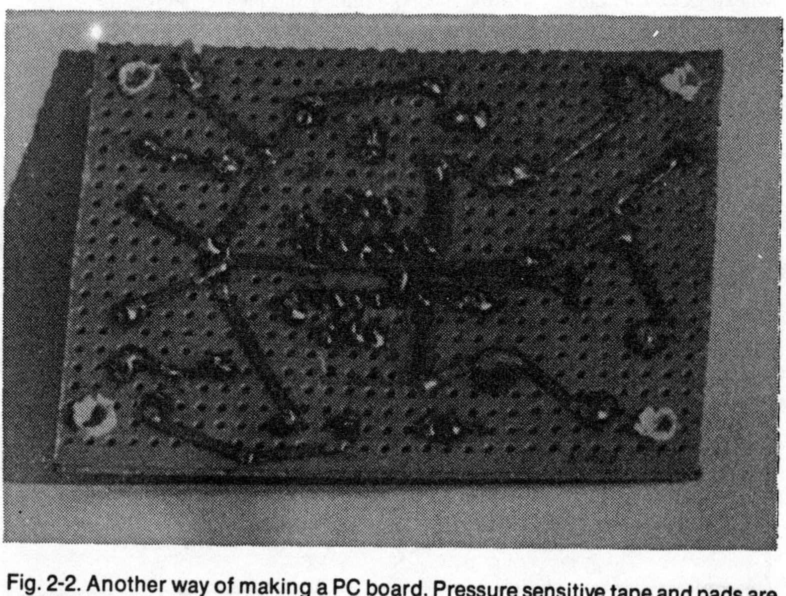

Fig. 2-2. Another way of making a PC board. Pressure sensitive tape and pads are laid down and soldered together. This system only works with lightweight components.

are several places where the lines could be greatly shortened to obtain a smaller board. The terminal points are usually terminal pins of one type or another soldered to the board. The pins make interconnections between the board and the chassis much easier. They should be labeled for ease in working. A tiny piece of gummed tape with "GRD", "V+", or "OUT" can be of help when interconnecting.

Before sketching your PC pattern, have all of the components on hand so you know exactly how much space they will require. One other way of reducing the size of a board is to mount the components vertically instead of the more usual way of lying horizontally on the board.

When drawing the pattern on paper, do you find two lines that cross but do not interconnect? Either employ a component to bridge the crossing, or use an insulated jumper wire to "leap" across the line.

First experiment with a simple circuit. There are several in the book to start you off. A two-transistor project with only a few associated components is the way to begin.

Although not strictly a printed circuit technique, an excellent way to achieve a high density board without a PC pattern is to use an uncoated perforated board. At the same time, buy a number of the pins which are made to fit into the fine holes that make the board resemble a mesh. The pattern of these holes makes it possible to push through the pins of an IC or its socket. Then point-to-point wiring is used from components to the pins and even using wire for interconnections.

All of these methods have this in common: They require care, accuracy, cleanliness, and the avoidance of too large a flow of solder. In the typical board where lines are close together, shorts happen too easily from molten solder making an unwanted connection. A magnifying glass comes in handy to examine the spread of the solder and the quality of the solder joint. The glass can be of help in detecting fine breaks in a line of a PC board.

One further point is worth mentioning. That is the use of sockets. They are useful for transistors but particularly for ICs. While it is true sockets can be

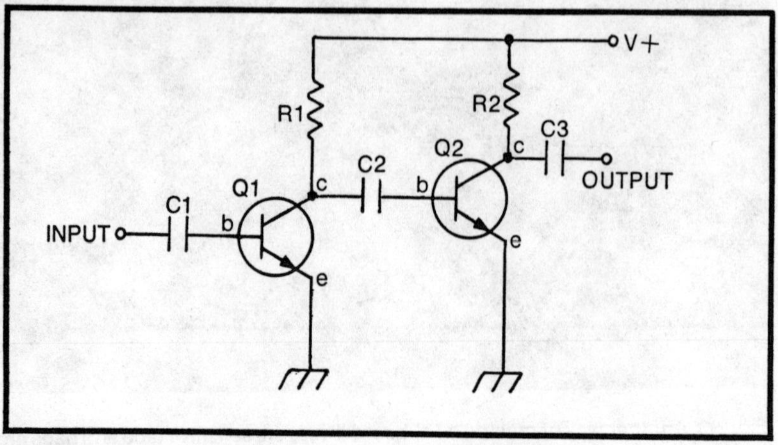

Fig. 2-3. A typical two-transistor schematic (needs bias resistors for proper operation) from which a printed circuit board is to be made.

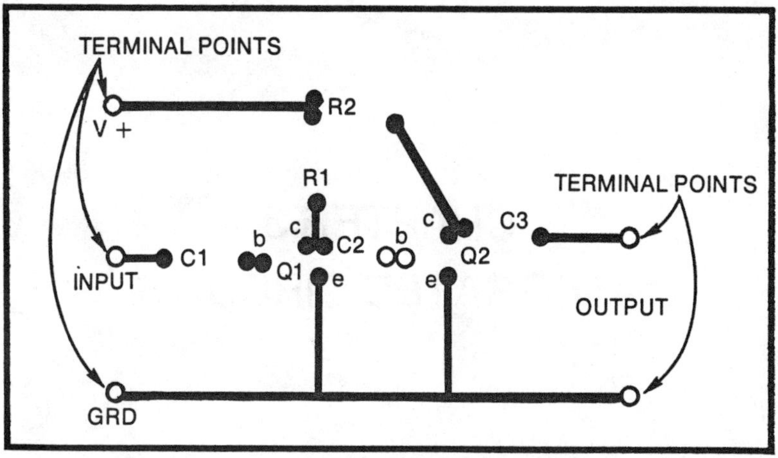

Fig. 2-4. Sketch for etching pattern of the circuit shown in Fig. 2-3. Although drawn to show the similarity to the original schematic, all of the connecting lines should be reduced and in some places even eliminated.

the cause of noise in a circuit as well as occasional malfunctions. they offer the advantage of allowing you to interchange different ICs without desoldering. Trying to desolder 12 or 14 pins usually leads to the destruction of the IC. even if it was not defective before.

When you are not certain that a particular transistor will work in a circuit. or are trying to find the best one for that purpose. a socket should definitely be used. When experimenting with a circuit. it is not uncommon to accidently reverse the voltage or doing something equally foolish. The socket then becomes a real time saver. Simply pull out the IC and insert another in its place.

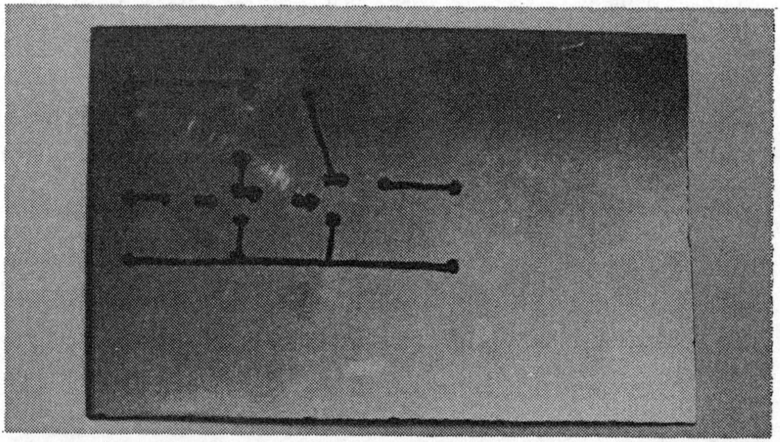

Fig. 2-5. Copper-clad board with the etch resist applied. Board is now ready for etching.

CHAPTER 3
INTEGRATED CIRCUITS

In spite of constant advances in electronics, vacuum tubes are not entirely obsolete and neither are transistors. But, integrated circuits have managed to take over many of the functions formerly associated with the older technologies. Without ICs, the new mini and micro computers would not exist. IC operational amplifiers internally combine quantities of transistors with their associated circuitry. Some ICs have FETs in the input stage so that you have a solid-state circuit whose impedance matches that of a vacuum tube. Some ICs have internal compensation, such as the 741. Others allow external compensation to tailor the IC to fit almost any need, the 709 being typical of that group. Some ICs are general types, such as the 709 and 748, which can be used for a variety of purposes. Others are highly specialized devices. The 555 is an example of a highly specialized IC. It is a timing device allowing on or off periods ranging from a microsecond to days. The 555 can also produce an excellent square wave over a wide range of frequencies with the addition of only a few discrete components. Still other ICs can generate two or even three different waveforms. One power supply listed in Chapter Four uses an IC as an output regulator.

With newer designs being created continuously, more and more "goodies" are packed in these tiny packages. Some are short-circuit proof; even output shorting will not damage some units. There are two and four input preamps. An example of the four input device is RCA's CA3048 with one output. The LM381, a product of National Semiconductor, has two non-inverting inputs, and two inverting inputs with two outputs. It is used in this book as a stereo preamp.

Variations are endless and more keep coming out on the market. So, it becomes a matter of finding the IC that meets the demands of that particular circuit.

One of the many advantages of integrated circuitry is that all the components are imbedded in the same package, so that temperature differences, such as might exist between two discrete transistors. do not cause drift. In the old days. i.e.. before ICs. 21 transistors plus some 27 or so resistors and capacitors took up a lot of room. Yet you can obtain all of these components in one IC. Motorola's HEP583 contains them all on one tiny chip the size of the nail on the little finger.

Furthermore. every interconnection in a circuit is a potential source of trouble. Now. one merely plugs in or solders 8 to 16 pins to place the whole package into the circuit.

Audio amplifiers using the newer ICs offer a wide frequency range. The 540. a product of Signetics. can be used as a power amp or as a preamp. It has low input noise and allows full power at frequencies as high as 20 kHz with low distortion.

With an IC on a breadboard. the experimenter can have a holiday. Varying the supply voltage will show the effects. sometimes unnoticeably. sometimes dramatically. of voltage changes. But be careful. No IC is indestructable. One should be especially careful of not reversing input voltage leads. The maximum voltage as indicated for an IC should never be surpassed. or even not equaled. Voltage meters may indicate less voltage than is actually in the circuit. So. it is best to run a few volts below the indicated maximum.

Theoretical equations are not included in this book. Instead. a variety of practical applications are offered. A variety of ICs will be used. In many cases the circuit values can be improved for your own use. The real experimenter will regard these circuits as a challenge—a jumping off place for learning the fascinating world of micro-electronics.

None of the above is meant to suggest that there are no disadvantages related to this wunderkind. But in general. the disadvantages are far outweighed by the advantages. Due to manufacturing differences (tolerances). characteristics between two devices of the same type can vary widely. But testing an op-amp requires more costly equipment than is owned by most hobbyists. So. if the IC is suspect. substitute another one. This is why a socket is recommended. A power supply with a current meter is a great help in finding out if the IC is at fault. It is obvious that if the circuit is wired correctly and no current is being drawn. the IC may be at fault. Such a method is too simplistic to be entirely accurate. So. make a substitution after checking to making sure the rest of the circuit is not the culprit.

Fortunately most ICs are relatively inexpensive. They will range from a quarter to a dollar. An expensive one. the LM381. sells for only $2.50 each.

When installing discrete components they should be placed close to the IC. This will prevent possible malfunctions. Another problem associated with ICs is their noise. This is why the type should be chosen with care when building a high-gain audio circuit. In this respect. lead length should be kept to a minimum to maintain stability and to avoid parasitic oscillations. Input leads should not be dressed too closely to other signal leads. But this is true for any audio circuit. even for those built with vacuum tubes.

OPERATIONAL AMPLIFIERS

The symbol for an op-amp is a triangle with its point facing the output. See Fig. 3-1. There are two inputs: the upper is the inverting input and is indicated with a minus sign next to the lead. The lower is the non-inverting input with a plus sign next to it. At the apex of the triangle is the output.

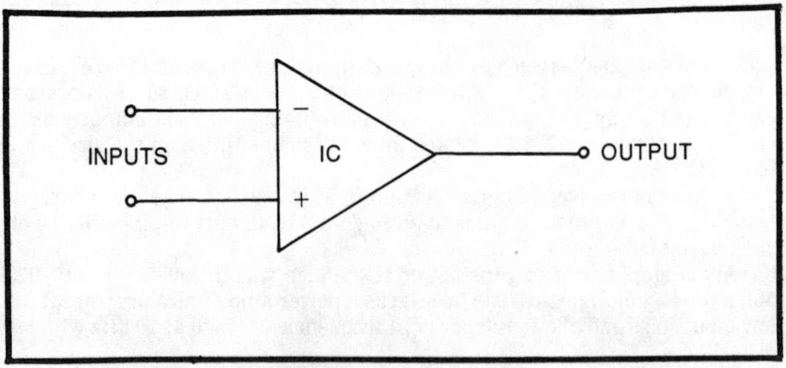

Fig. 3-1. Symbol for op-amp.

Some op-amps require a single-ended power supply: positive and ground. Others demand a positive and negative voltage as well as ground. Many ICs are extremely tolerant of the level of input voltage. One unit requiring a maximum of 30 volts was found to operate quite well with as little as nine volts.

Pin numbering should be checked carefully before wiring. Some ICs have a dot indicating pin 1. but not all. So. use the appendix in the back of the book which lists the pin numbers for various ICs.

The following chapters offer examples of ICs at work. Transistors. as was pointed out. are not obsolete. so there will also be samples of their use. Often a transistor is simpler to use: the IC being too complex for that purpose. So. a single transistor with a couple of resistors may be adequate.

DIGITAL ICs

So far we have been speaking mostly of op-amps. But the field of ICs covers not only linear ICs. of which the op-amps are characteristic. but there are also the digital ICs. Symbols for three types which you are likely to encounter are shown in Fig. 3-2. HEP584. the IC used in the Party Lie Detector project. is a dual input gate. but is not drawn as shown in the symbol. A circle is used instead. as it makes the schematic easier to follow. The 555. which uses a section of its own. is a flip-flop and thus is shown as a rectangle.

This is only a very sketchy outline of the field of ICs. For the experimenter who wishes to go further he should consult a textbook on the subject. such as TAB Books No. 771. Integrated Circuits Guidebook. or No. 787. Op Amp Circuit Design & Applications.

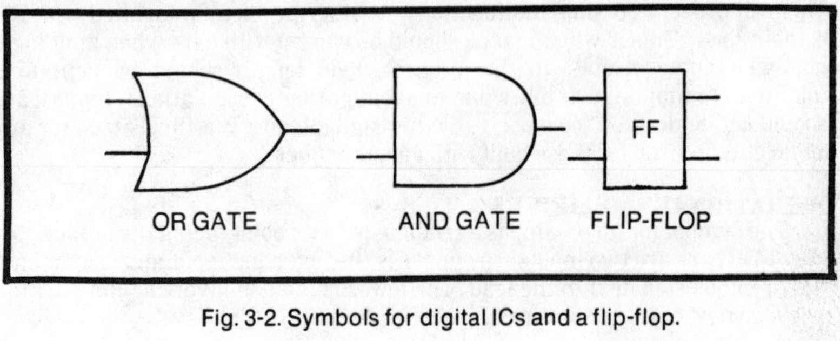

Fig. 3-2. Symbols for digital ICs and a flip-flop.

CHAPTER 4
POWER SUPPLIES

One major component common to all active circuits is the power supply. It can range from a 1.5-volt dry cell for powering a flashlight to the complex laboratory type, fully regulated, short-circuit proof, current and voltage limiting. Some supplies must furnish both positive and negative voltages for certain ICs, and which must track within fractions of a volt of each other. However, not every electronic supply has to be that sophisticated. It depends entirely on what the circuit needs to operate successfully. You don't use an elephant gun to kill a mosquito. This brings us to the reason for the different supplied described in this chapter.

A portable project with a low current demand will need only one or more dry cells. Once the voltage or current needed goes up, so that the cells would be used up too quickly, then an electronic supply becomes necessary.

Dry cells have the advantage of being simple: hum is never a problem and they keep the circuit small and lightweight. No big capacitors, transformers, zeners, or rectifier diodes are needed. The whole project can then be encased in a smaller chassis (as well as being cheaper) since fewer parts are needed. A battery is a must if the unit is to be portable, of course.

Sometimes the current demand is low, but you are tired of buying batteries, especially when the unit is used at home. And batteries do end up being quite costly in the long run. Although making up a battery eliminator may seem more expensive at first, it does not take many hours of operation before its cost has been amortized.

A SIMPLE BATTERY ELIMINATOR

The project illustrated in Fig. 4-1 will furnish a nominal nine volts with enough current to power any unit that ordinarily employs a 9-volt battery

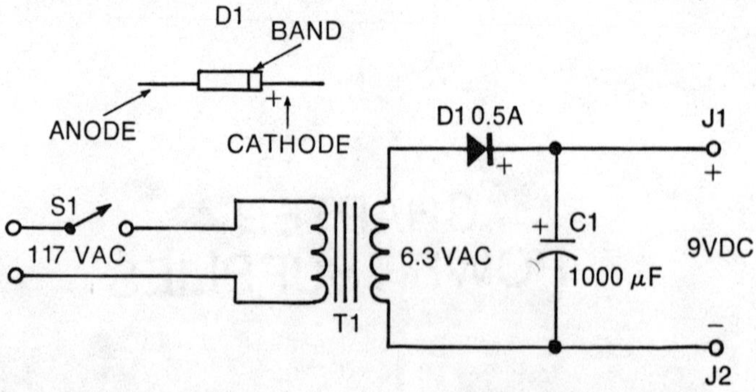

PARTS LIST

C1—1000 μF, 35-volt electrolytic capacitor (Radio Shack 272-1019, or equal)

D1—50-volt (PIV), 1 ampere rectifier diode (Radio Shack 276-1101, or equal)

J1, J2—Terminal strip or binding posts as desired, to match unit to be powered.

S1—Single-pole, single-throw switch

T1—6.3 VAC, .5 ampere transformer (Olson Electronics TF-047, or equal)

Fig. 4-1. Schematic for a 9-volt battery eliminator. This supply can be used to operate transistor radios, tape recorders, etc.

The Circuit

The circuit is a simple one. The transformer (T1) drops the 117 volts to a low 6.3 volts. The diode (D1) converts this from AC to pulsating DC. The 1000 μF capacitor (C1) smoothes out the supply output. A note about rectifying diodes: their rating should be at least twice that of the voltage being handled. The diode can be as large as you can find, in terms of voltage and current carrying capabilities. Too large does no harm, but too small—poof, no diode is left after the power is turned on. Very little current is consumed when the radio is not playing, so even if you leave the power supply switch on your light bill will not go up appreciably. J1 and J2 can be any type of connectors that will fit the unit you plan to use with the power supply.

A DUAL VOLTAGE SUPPLY

The battery eliminator just described is adequate for many purposes. However, it suffers from two faults: (1) it offers only one output voltage, (2) it is not short-circuit proof. There are times when we would like to have one power supply which is able to furnish two of the most popular voltages—6 and 9 volts. So instead of building two power supplies, the circuit of Fig. 4-2 offers either of two voltages with the flick of a switch. Furthermore, both voltages are regulated and protected against accidental shorts. Let's see how.

In Fig. 4-2, transistor Q1 remains in a non-conducting stage with no load. With an applied load, the current increases until the voltage at the output is slightly lower than the zener's voltage. In position 1 of switch S2, this voltage would be determined by Z1. If the voltage increases beyond that point, it will

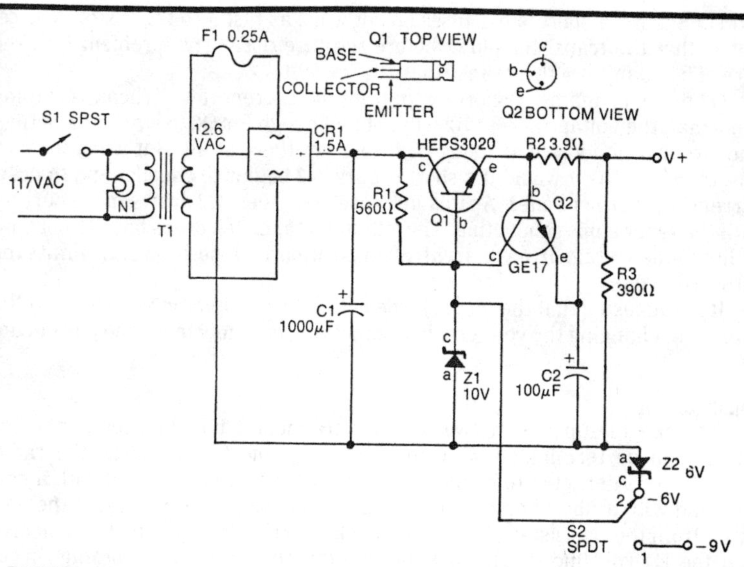

PARTS LIST

C1—1000 μF, 35-volt electrolytic capacitor (Radio Shack 272-1019, or equal)
C2—100 μF, 35-volt electrolytic capacitor (Radio Shack 276-1016, or equal)
CR1—50-volt (PIV), 1.5 ampere bridge rectifier (Radio Shack 276-1151, or equal)
F1—.25 ampere fuse
N1—Neon pilot lamp (Radio Shack 272-703 or equal)
Q1—HEPS3020 power amplifier transistor (Radio Shack 276-2018, or equal)
Q2—GE17 silicon power transistor (Radio Shack 276-2009, or equal)
R1—560 ohm, 1/2 watt carbon resistor
R2—3.9 ohm, 1/2 watt carbon resistor
R3—390 ohm, 1/2 watt carbon resistor
S1—Single-pole, single-throw switch
S2—Single-pole, double-throw switch
T1—12.6 VAC, 1 ampere transformer (Olson Electronics TT-076, or equal)
Z1—10-volt zener diode, 1/2 watt (1N5240B, or equal)
Z2—6-volt zener diode, 1/2 watt (1N5233B, or equal)
Misc.—Case, 5×2-3/4×2 inches

Fig. 4-2. Schematic of the dual supply which is both regulated and short-circuit proof. It will furnish up to 1 ampere.

cause the voltage across the base-emitter junction of Q1 to drop. This will cause the collector current to drop also. But if the output voltage drops, the voltage across the base-emitter junction will rise, causing the collector voltage to rise. This see-saw maintains the output at a constant level. Q1 alone will not protect against a short, since fuses do not work as fast as a transistor. The result is that the transistor blows before the fuse does. This problem is taken care of by adding another transistor to the circuit.

Q2 protects against a short by limiting the current flow. When everything is normal, the voltage across R2 will not be enough for Q2 to start conducting. The circuit will operate as though Q2 were not there at all. But when the current increases, as it would if a short occurs, Q2 begins to conduct and to draw current from either zener. As this increase continues, Q2 draws more current until the zener stops conducting. The output voltage then drops back to normal. Z2 limits the voltage to 6 volts with S2 in position 1; in position 2, Z1 limits the output to 9 volts.

It is suggested that the unit is turned off before switching from 6 to 9 volts, to prevent changing the voltage when something is connected to the power supply.

Construction

The panel lamp and the two switches are mounted on the front panel. The output can be terminated with any type of connector to match the radio, cassette recorder, etc., to be powered. Q1 should best be mounted with a heat sink and with a liberal application of silicone grease to help conduct the heat away from the transistor. There is no tricky wiring. It should be remembered that the 117-volt line should pass through the chassis with a grommet in the hole to avoid fraying the cord. Figure 4-3 is a photograph of the unit. A phono jack was used for the output since the current passing through it was small.

AN UNREGULATED POWER SUPPLY

For our next example we have a project that draws 100 milliamps at 12 volts. Looking at mail order catalogs, or in your local electronic store, you will

Fig. 4-3. A dual voltage supply. This unit furnishes both 6 and 9 volts.

see a transformer rated at 12.6 volts and capable of supplying one-half ampere (500 milliamps). We need four diodes, each rated at about twice the voltage they will be handling. We will choose diodes rated at 50 Peak-Inverse-Volts (PIV). They rectify the voltage, changing the AC to pulsating DC. This charges the capacitors. The capacitors should range from 200 to 1000 μF; the larger capable of furnishing more current and reducing the amount of ripple in the final stage. The capacitors should be rated at about twice the voltage used. The 1-ohm resistors are added to reduce the current surge when the power is turned on. The supply now looks like Fig. 4-4.

Now to calculate the regulation needed. The most simple way to do it is to add a resistor and a zener diode.

A REGULATED POWER SUPPLY

Designing a regulated power supply to meet your particular wants is not difficult. With only pencil, paper and a working knowledge of Ohm's law you can formulate a supply design for any specific project. In the example that follows we will design a 12-volt supply. Once you understand the principles, you can apply them to build a unit supplying any particular voltage and current.

We will begin with an unregulated supply able to provide at least the amount of current your project will draw and a voltage higher than needed (about 30% more to allow for the drop due to regulation).

What is the difference between an unregulated and regulated supply? The unregulated will furnish, according to our example, 12 volts. However, as the current demand increases, the voltage decreases. This means that in operation you may have a supply which has dropped down to nine or ten volts. For a

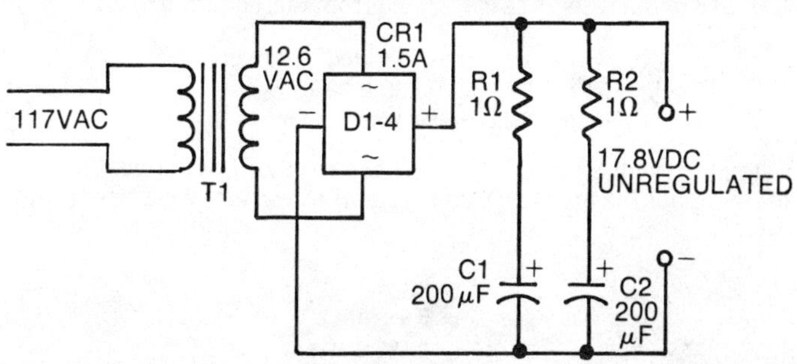

PARTS LIST

C1, C2—200 μF to 1000 μF (see text), 35-volt electrolytic capacitor

CR1—50-volt (PIV), 1.5 ampere bridge rectifier (Radio Shack 276-1151, or equal)

R1, R2—1 ohm, 1 watt carbon resistor

T1—12.6 VAC, .5 ampere transformer (Olson Electronics TF-047, or equal)

Fig. 4-4. A typical unregulated power supply. This type of supply can be employed wherever there is no objection to the voltage dropping when the current goes up.

critical circuit this drop will cause a malfunction. So, what is done is to feed the unregulated supply to a regulator and for the price of a few volts drop, which was calculated and allowed for, we get a 12-volt supply whose output does not change regardless of the current demand until, of course, the demand exceeds the capability of the diodes and transformer.

The circuit to be added—regardless of the final voltage required—follows the same principle. Rx and Zx must be calculated for your particular needs. The zener's job is to control the current across the resistor so that the current remains constant with maximum load or none at all. It should not be forgotten that the ordinary power supply without a load can go fffft! If the current remains constant, so will the voltage drop across the resistor.

First choose a zener rated at the voltage required. In our case, we need a 12-volt zener. The wattage of the zener is calculated from the formula:

$$Px = Vz\,(I\,max - I\,min)\,1.20$$

Px is the necessary wattage
Vz is the voltage of the zener—in our case it is 12.
I max is the maximum current which will be drawn, or 100 mA.
I min is the minimum current: without a load it is zero.
The number 1.20 represents a safety factor to provide adequate current for correct regulation.
Substituting numbers in the formula we get:

$$Px = 12\,(100)\,1.20 = 1.44 \text{ watts}$$

Therefore, a 2-watt zener is the answer to the requirement of this power supply.

The value of the resistor is calculated from this equation:

$$\begin{aligned} Rx &= (Vin - Vout) \text{ divided by } I\,max + 20\% \\ &= (17.8 - 12)/(100 + 20\%) \\ &= 5.8/0.12 = 48 \text{ ohms} \end{aligned}$$

A standard value for a 10% tolerance resistor is 47 ohms. Close enough Now to figure out the tolerance wattage of the resistor. We know from Ohm's law that:

$$\begin{aligned} P &= EI \\ &= 5.8 \times 0.12 \\ &= 0.696 \text{ watts} \end{aligned}$$

Therefore, 1-watt resistor should be used.
Figure 4-5 is the schematic for the regulator circuit.

SOLID STATE REGULATOR

A solid state regulator in a 3-terminal package is currently available from several manufacturers including Radio Shack, Poly Paks, and several others. They are available in ranges from 5 to 24 volts and with current ratings up to one ampere. You merely feed a slightly larger unregulated voltage to the regulator which then accomplishes what we were doing with the zener and resistor. However, the latter combination allows you to calculate exactly what you want. Sometimes one wants an odd-ball voltage which is not available us-

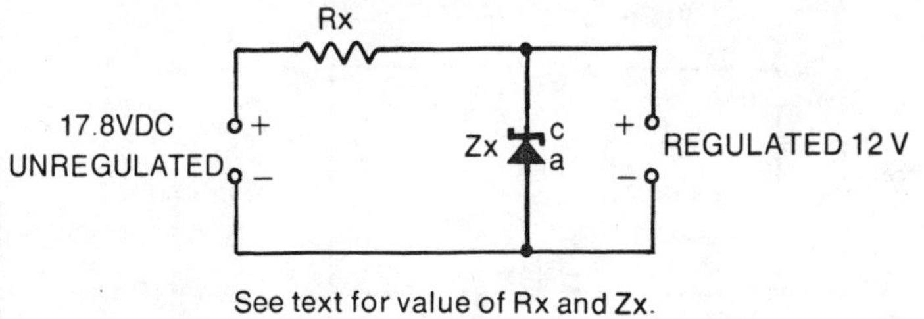

Fig. 4-5. Zener-resistor combination to provide regulation.

ing a packaged regulator. Figure 4-6 shows the regulator in a typical circuit of a power supply.

5 TO 20 VOLTS AT 1 AMPERE

Although a solid state regulator is designed to offer a fixed output voltage, it can be re-designed into a circuit so that an adjustable supply results. Figure 4-7 shows the LM309K, nominally a 5-volt regulator, which can be manipulated to offer a continuously adjustable voltage output from 5 to 20 volts at a current level of up to 1 ampere. The circuit is simple and easy to build. It is equal in performance to many power supply circuits boasting of countless components and very complex.

The regulator at pins 1 and 2 sees an unregulated voltage. The potentiometer (R2) presents a variable voltage within the limits of the circuit. Even the operation of the supply without a load does not affect its regulation. This is a unique adaptation of a fixed regulator. The two meters, one for reading voltage and the other for current, are nice to have but are dispensable. They

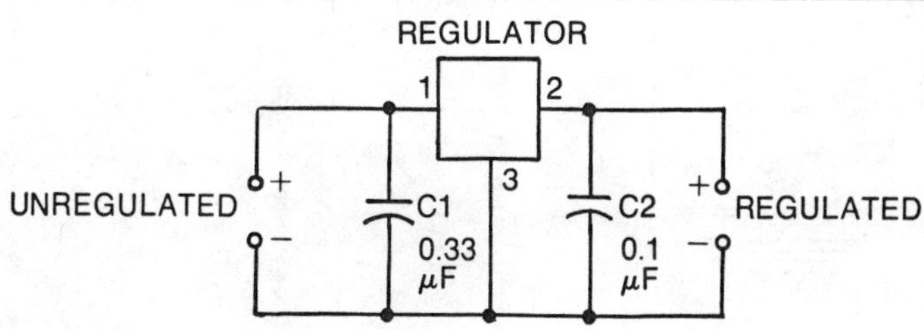

PARTS LIST

C1—0.33 μF MYLAR TUBULAR CAPACITOR

C2—0.1μF MYLAR TUBULAR CAPACITOR

REGULATOR—TO FIT DEMANDS OF CIRCUIT (RADIO SHACK 276-1770, 5V at 1 AMP; 276-1171 12V at 1 AMP, 276-1772, 15V at 1 AMP OR EQUAL)

Fig. 4-6. A solid state regulator.

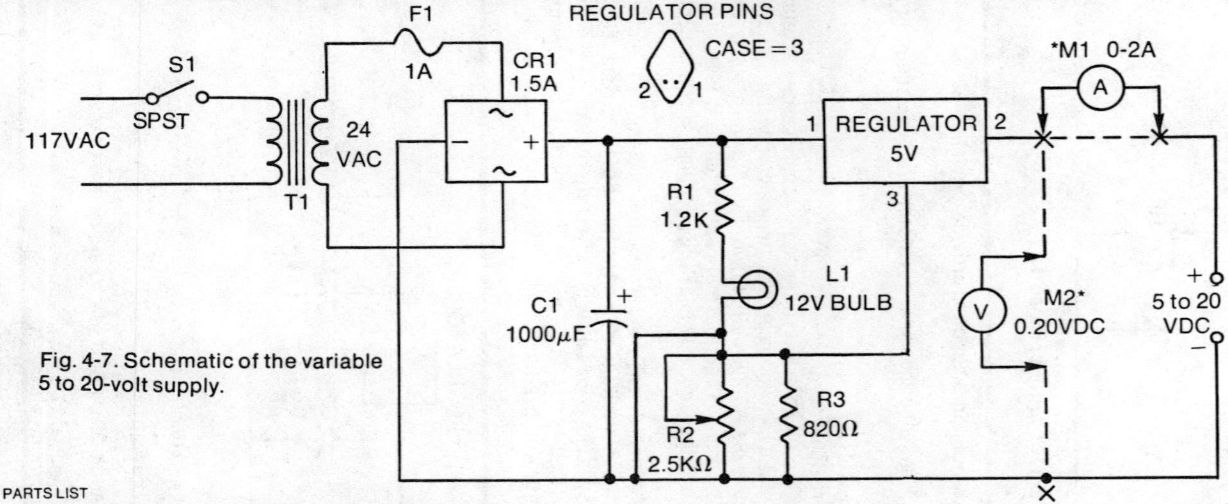

Fig. 4-7. Schematic of the variable 5 to 20-volt supply.

PARTS LIST

C1—1000 μF, 50-volt electrolytic capacitor (Radio Shack 272-1047, or equal)
CR1—100-volt (PIV), 1.5 ampere bridge rectifier (Radio Shack 276-1152, or equal)
F—1 ampere slo-blo fuse
J1—Red 5-way binding post
J2—Black 5-way binding post
L1—12-volt pilot lamp
M1—0-2 ampere DC ammeter
M2—0-20 VDC voltmeter
R1—1.2K, 1 watt carbon resistor
R2—2500 ohm potentiometer with single-pole, single-throw switch
R3—820 ohm, 1 watt carbon resistor
Regulator—LM309K 5-volt, 1 ampere voltage regulator (Radio Shack 276-1770, or equal)
S1—Single-pole, single-throw switch (mounted on R2)
T1—24VAC, 2 ampere transformer (Radio Shack 273-1512, or equal)

Complete parts kit is available from:

RAMCO Electronics
365 Canal Street
New York, N. Y. 10013

can be eliminated by using a VOM or other voltage meter. Once the voltage is set, the current can then be read by the same meter by changing its position. Another way would be to calibrate the pot with a known voltage meter. Once it is done, no further calibration is necessary.

There are so few parts that no board is needed for wiring. Connections can be made to an output terminal strip to which the bridge rectifier is soldered.

The regulator uses the chassis on which to rest but it should be insulated using mica washers. A large heat sink is an absolute necessity. A chassis measuring about 7 × 5 × 2 inches is plenty large enough. The transformer is also mounted on top of the chassis. See Fig. 4-8.

This supply really belongs on the active hobbyist's workbench due to its current carrying capability and the fact that its voltage range encompasses most needed voltages.

MOBILE POWER SUPPLY

A somewhat different power supply is one designed to operate from a car or boat. When camping it allows you to operate cassette recorders, portable radios, and such. The unit is plugged into the cigarette lighter jack of the car. From the car's 12-volt negative ground system you can obtain, by means of the circuit illustrated in Fig. 4-9, three different voltages. These are: 6, 7.5, and 9 volts DC. The zeners combined with the series transistor can power any unit which does not demand more than 300 milliamperes.

The explanation of the circuit is identical to that of Fig. 4-2. There are two major differences: three instead of two selected voltages, and it is powered from a 12-volt system.

This circuit is also short-circuit proof as well as being well regulated.

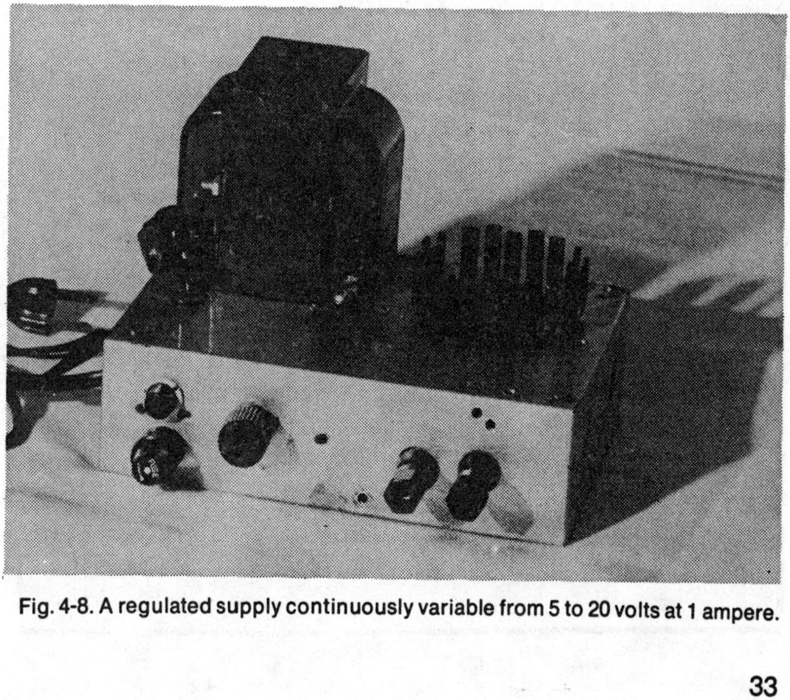

Fig. 4-8. A regulated supply continuously variable from 5 to 20 volts at 1 ampere.

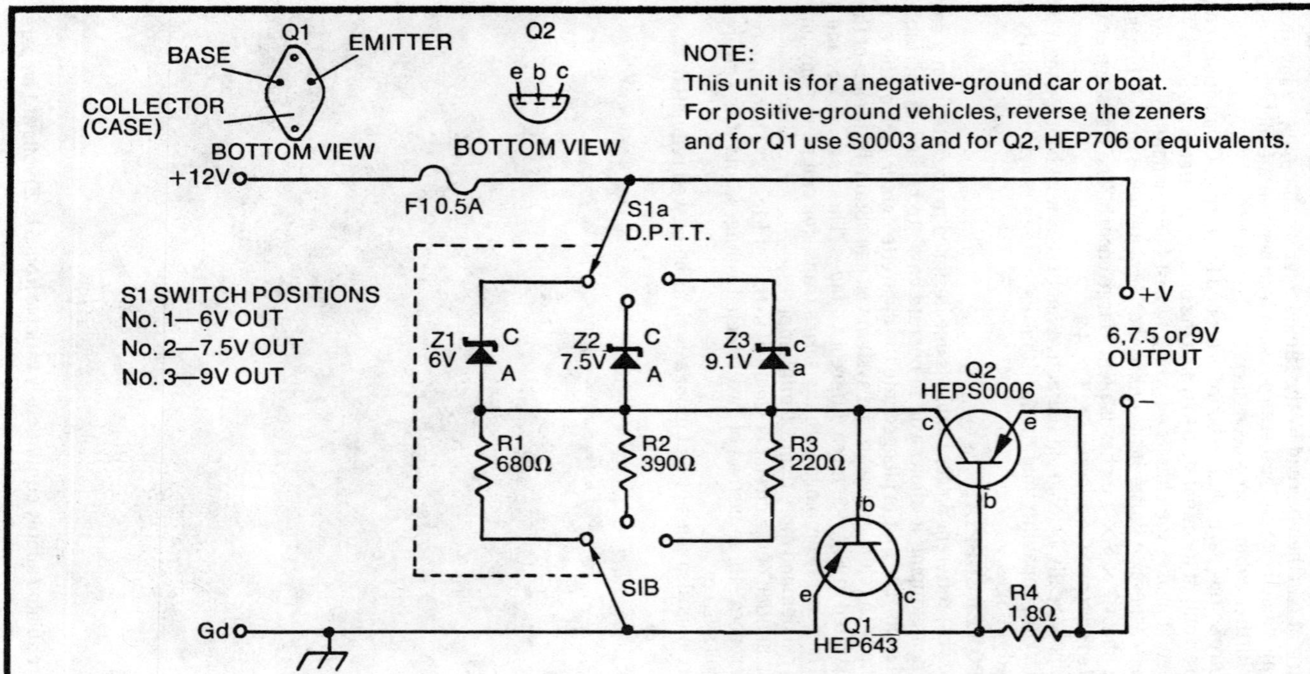

PARTS LIST
F1—.5 ampere fuse
Q1—HEP643 power amplifier transistor (Radio Shack 276-2006, or equal)
Q2—HEPS0006 amplifier transistor (Radio Shack 276-2034, or equal)
R1—680 ohm, 1/4 watt carbon resistor, 10% tolerance
R2—390 ohm, 1/4 watt carbon resistor, 10% tolerance
R3—220 ohm, 1/4 watt carbon resistor, 10% tolerance
R4—1.8 ohm, 1/4 watt carbon resistor, 10% tolerance
S1—Double-pole, triple-throw switch (Olsen Electronics SW-846, or equal)
Z1—6-volt, .5 ampere zener (1N5340B, or equal)
Z2—7.5-volt, .5 ampere zener (1N5343B, or equal)
Z3—9.1-volt, .5 ampere zener (1N5346B, or equal)

Fig. 4-9. A three voltage mobile supply designed to operate from cars, campers, and boats.

Some of the older European cars have a 12-volt positive ground. In that case. PNP transistors would have to be used. The capacitors would have to be reversed and so would the zeners. in regards to polarity.

VARIABLE AC POWER SUPPLY

So far we have been dealing only with DC power supplies. But. there are many times when an AC supply becomes very handy. And. the variable AC supply described below belongs on every experimenter's workbench. Refer to Fig. 4-10. Such a supply is erroneously called a Variac. Erroroneously because the name is the property of General Radio and is not a generic name. Its real name is an autotransformer. But. whatever its name. it will supply a continuously variable AC voltage from 0 to 120 or 140 volts. Some of these transformers are able to supply up to five amperes.

The AC supply has many uses. For example. by adding a fixed DC supply to the output of the autotransformer. you can control your DC output from zero to its maximum output voltage. In actuality. its most valuable use is for testing a new circuit. Instead of plugging in your new circuit through its power supply directly to the power line and getting smoke for your pains. try this instead: Plug the test circuit's line input into the autotransformer. Turn it on and gradually turn up the AC power as you very carefully watch the voltage and current readings. If the current jumps up suddenly when only a fraction of the voltage has been turned on. you have something wrong. Turn off the power at once. The number of semiconductors you will save over a period of time will quickly make up for the cost of the autotransformer. The trouble with a fuse is that it does not cut off fast enough to protect transistors and ICs. So use your autotransformer with every new circuit that can be plugged into the 117-volt line.

Fig. 4-10. A variable AC supply.

Another use for the unit is as a substitute secondary for a power supply when you do not have the right transformer and you are anxious to try out a new circuit.

Still another use: one evening the author's soldering pencil stopped working and only the 100-watt gun was left to use. Much too high a wattage for soldering on a PC board. By plugging the iron into the autotransformer and furnishing much less voltage. the iron became a low wattage unit. Not the most effecient in the world. but better than taking a chance of ruining a copper clad board or waiting till the next day.

According to Ohm's law, a 100-watt iron consumes almost 0.9 amperes of current at 110/120 volts.

The question is. now how much wattage will be gotten if the voltage is cut to 20 volts? Using Ohm's law again we have:

$$P = E \times I$$
$$= 20 \times 0.9$$
$$= 18 \text{ watts}$$

Since up to 30 watts is safe for a PC board. that high powered gun is now safe to use. It will take longer to heat up. but it will work safely. Many more uses will suggest themselves to you as you go along your experimenting path.

Construction

Since you are dealing with high currents, wiring should be done carefully with heavy gauge wire. Make certain every solder joint is solid. The unit used as an illustration in Fig. 4-11 is able to carry three amperes. So. it is fused to limit any current above that. which could only be caused by a short in your wiring of the autotransformer or in the circuit being tested. Also. all of the switches must be rated to handle five amps for safety. The shunt for the current meter also has to be calculated. based on the internal resistance of that meter and cannot be expressed in the circuit diagram. The author used Nichrome wire for the shunt. This is used as a heating coil in electric heating appliances. It can be purchased from most hardware stores. A store that repairs such appliances might just give you two to three inches of the wire. as that is all you will need.

Attach one end of the wire to one side of the meter switch (HI position terminal) and the other by means of an alligator clip to point A in the diagram. Use a 100-ohm resistor, rated at least at 50 watts, as a load. Do not leave the power on any longer than necessary as the resistor will get very hot and will eventually melt. Put S2 in its LO position. Turn up the voltage until you read 100 volts on M2. M1 should now read full scale, or one ampere. Switch to the HI position of the current meter and it should now read somewhere around one-third scale, or 1 amp. If it reads too low, you have too much resistance and the Nichrome length should be reduced; too high, not enough. Turn off the power. Not cutting down on any of the measured length, wrap the Nichrome around the terminal at A, since Nichrome cannot be soldered. Let the load resistor cool and try again. The meter reading should be the same as before. Now cut off any excess Nichrome. Of course, if you or a friend has an accurate current meter, you can use that as a standard.

VOLTAGE DOUBLER

The circuit described here multiplies the AC input voltage and then rectifies it to DC. See Fig. 4-12. Although none of the components have been given

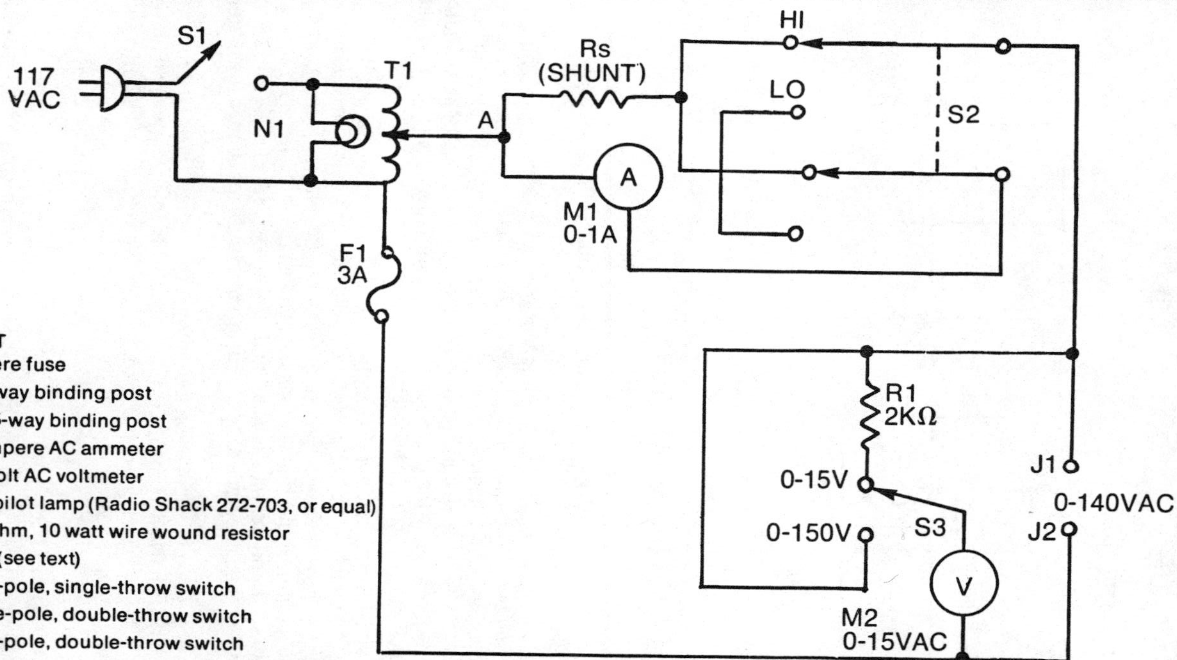

PARTS LIST
F1—3 ampere fuse
J1—Red 5-way binding post
J2—Black 5-way binding post
M1—0-1 ampere AC ammeter
M2—0-15 volt AC voltmeter
N1—Neon pilot lamp (Radio Shack 272-703, or equal)
R1—2000 ohm, 10 watt wire wound resistor
Rs—Shunt (see text)
S1—Single-pole, single-throw switch
S2—Double-pole, double-throw switch
S3—Single-pole, double-throw switch
T1—Autotransformer (Olson Electronics TT-279, or equal)
Misc.—Case, 12×4×7 inches

Fig. 4-11. Schematic for the AC supply ranging from 0 to 140 volts at 1 ampere.

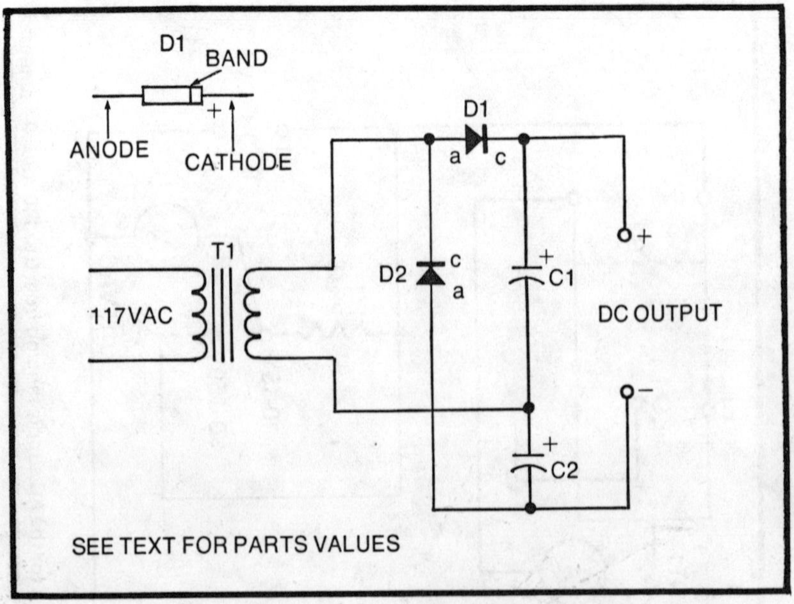

Fig. 4-12. Example of an unregulated voltage doubler.

values. they are simple to figure out to suit your need. First. decide how much voltage you want to end up with. Divide that by half. and that gives you the voltage rating of the transformer. Its current rating is dependent on the circuit. Always be generous in that direction. Too much current available is not bad. while running the transformer close to its upper current limit makes it run hot and may eventually burn it out.

D1 and D2 must have a PIV rating at least twice that of the DC output voltage. The current rating depends on what the following circuit demands. Don't be skimpy in choosing these current ratings either. It never does any harm to obtain a much larger diode in terms of its current rating. The capacitors. C1 and C2. should be a minimum of 500 μF with a working voltage twice that of the DC output voltage. The greater the capacitance the better the filtering.

The transformer supplies an AC voltage. On the positive half cycle of that voltage. C1 is charged by means of the diode D1. On the negative half of the AC voltage. C2 is charged through D2. Adding the two voltages together results in the DC voltage doubled.

VOLTAGE TRIPLER

There are times when you may need much more voltage than you have available from a simple power supply. However. you may not have a transformer which can boost the line voltage up to where you need it. This circuit (Fig. 4-13.) will give you almost three times the output voltage of the secondary of your transformer.

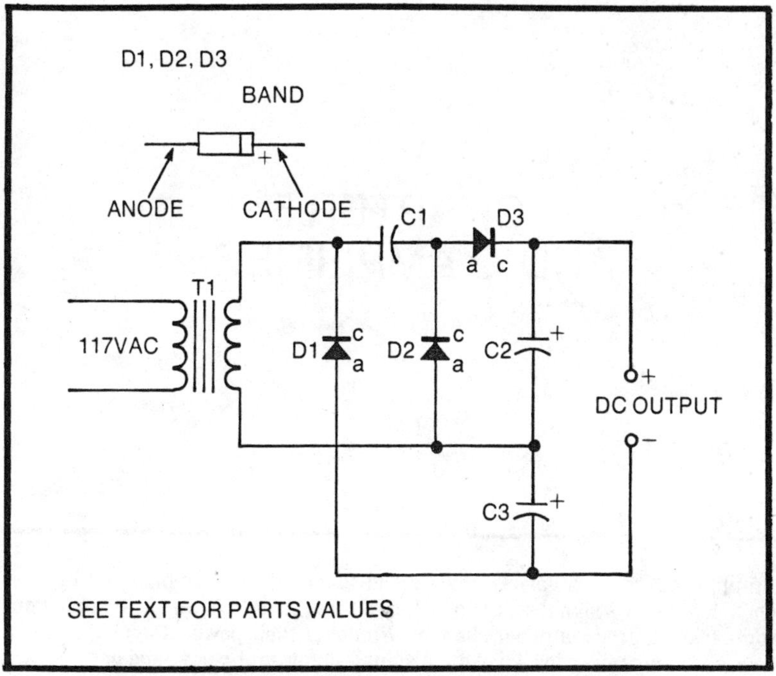

Fig. 4-13. Typical circuit of an unregulated voltage tripler.

Let us say you have a 24-volt transformer. This is a popular unit. With this circuit, you will obtain about 70 volts DC. As with any power supply, when you are dealing with line voltage, make certain rubber grommets surround the holes in the metal chassis through which the wires will go.

Divide by three the desired end voltage to calculate the transformer you need. If you cannot find a transformer which comes out exactly to the wanted voltage, get one a little bigger. Then, drop that voltage at the other end of the supply with a voltage regulator or zener-resistor combination. The values of the diodes are calculated just as in the previous project. They all have the same value.

C1 is charged through D2 with the negative half cycles of the AC voltage. C3 is charged through D1 at the same time. But, on the positive half cycles, C2 is charged through D3 to which is added the full AC voltage of the secondary of the transformer. The resulting output voltage is the sum of that across C2, the secondary AC, plus the charge across C1, and that from C3. This totals almost three times that of the original secondary voltage. It does not quite add up to exactly three times, since there are voltage drops through the diodes, but it is very close.

CHAPTER 5
AUDIO AMPLIFIERS

Amplifiers come in a variety of sizes and shapes. Their size may be tiny, just enough to boost a signal so it can be heard through earphones, or it can put out some two hundred watts per channel. Whatever their power, they have this in common: they take a low level input signal—a few millivolts—and boost it by a large factor to be able to move the diaphragm of a loudspeaker.

Ideally, the amplifier should raise the power level of a signal without introducing any sounds of its own. It will naturally amplify any noise that is inherent in the low level signal and which will be amplified a stage or two in the preamp before reaching the final stages of amplification. To build an amplifier that satisfies our ears means care in putting it together, making leads as short as possible, and above all, making certain that no hum originates in the power supply. That is the advantage of using a battery to power a unit. But unfortunately, batteries cannot be employed to power the bigger amplifiers. In that case, a well shielded transformer is a must. The power supply components must be located as far away as possible from the main section of the amplifier circuits. The metal case should offer good shielding against picking up stray signals from the air. A cover for the case must be provided for. With these few precautions, you are ready to tackle one or more of the amplifiers described in this chapter.

200 MILLIWATT AMPLIFIER

Should you need a small amplifier for a pair of earphones or for a simple intercom system, the circuit shown in Fig. 5-1 will fit the bill nicely. It can be put together with junk-box parts, or at the most, the whole circuit will entail few purchases. The output signal is fairly high, depending on the value of R2. It will never replace a high quality amplifier, but it is surprisingly good, considering the simplicity of the circuit.

Two transistors, Q1 and Q2, are used as a Darlington pair. The two collectors are connected together, but Q1 acts as a variable bias resistor for Q2.

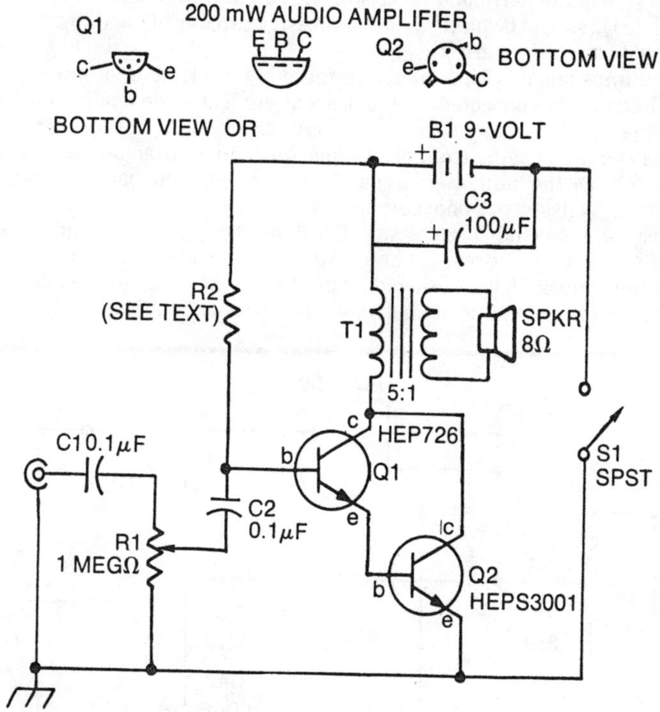

PARTS LIST
B1—9-volt transistor radio battery
C1, C2—0.1 μF mylar tubular capacitor
C3—100 μF, 15-volt, electrolytic capacitor
Q1—HEP726 general purpose silicon transistor (Radio Shack 276-2016, or equal)
Q2—HEPS3001 silicon power transistor (Radio Shack 276-2008, or equal)
R1—1 megohm, audio taper, miniature volume control
R2—See text
S1—Single-pole, single-throw switch
SPKR—8 ohm speaker
T1—5:1 ratio interstage transformer

Fig. 5-1. Circuit for a transistorized 200 mW amplifier.

Since the output is dependent on the input signal. the gain becomes approximately that of the product of the gains of the two transistors. This will range from 8000 to about 2 1/2 times that. Output of the amplifier is some 200 milliwatts. Not exactly wall shattering. but sufficient for its purpose. Various combinations of transistors were tried and most worked well with different gains. The overall gains varied. as was to be expected. dependent on the individual gain of the particular transistor. Q1 can be any general purpose audio type. while Q2 should be an RF power transistor capable of handling 200 mW of quiescent current. Both are NPN types.

Input impedance is high. therefore there is little loading of any circuit to which this amp is connected. The quiescent current is dependent on R2. To choose the optimum value based on your choice of transistors. substitute various resistors ranging from one to four megohms. Although an interstage transformer at the amplifier output is suggested. you can eliminate the transformer by using an 8-ohm earphone.

Construction couldn't be easier. Most of the parts are soldered to a terminal strip with the rest of the components mounted directly on the pot (R1). A fairly husky battery should be used for power. due to the current demand. or use the schematic in Chapter 4. the Battery Eliminator.

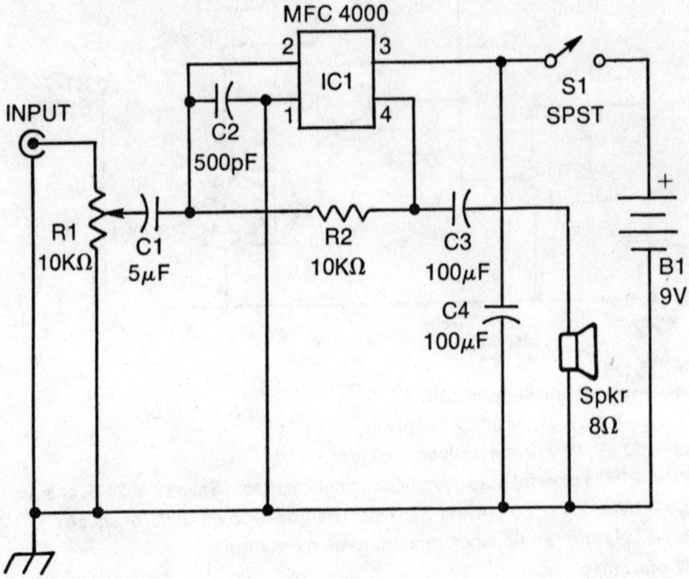

PARTS LIST

B1—9-volt transistor radio battery
C1—5 μF, 15-volt electrolytic capacitor
C2—500 pF silvered mica capacitor
C3, C4—100 μF, 15-volt electrolytic capacitor
IC1—MFC4000 (Motorola)
R1—10,000 ohm audio taper potentiometer
R2—10,000 ohm, 1/4 watt, low-noise metal film resistor, 10% tolerance
S1—Single-pole, single-throw switch
SPKR—8 ohm speaker

Fig. 5-2. Circuit for a 250 mW amplifier using an IC.

QUARTER WATT IC AMPLIFIER

The micro-power unit illustrated in Fig. 5-2 is an ideal amplifier for audio monitoring purposes. The big difference between it and the previous amplifier, apart from the extra 50 mW of output power, is that instead of a pair of transistors amplifying the original signal, one linear IC is used instead. The current demand is low and a battery will be long lived.

To study the advantages of one unit over the other, you might breadboard both and see for yourself just what the IC can do and what the transistors offer.

Its total output is 250 mW, low indeed, but it uses only the one IC and six other components. Input impedance is in the order of 10K ohms. Two such units are needed for stereo, but both will fit into a tiny box. Obtain a small speaker, two inches or so, if you want to mount it inside the chassis. Again, you will need two such speakers for stereo. The box can be all metal, or you can choose one with a plastic cover which is easier to drill. A hot pick or screw driver heated over the gas stove does the plastic drilling job. Drill a series of small holes for the speaker, one for the input jack, and a hole for the shaft of the volume control.

ONE WATT AMPLIFIER

This little amplifier is meant for low power applications with an 8-ohm speaker (or even into a lower impedance speaker). The heart of the unit is an IC. The 540, make by Signetics, is a 10-pin op-amp which will deliver one watt with a total harmonic distortion of less than 1.0%. The 540 offers short circuit protection as long as the maximum dissipation point is not reached. It requires low standby current with typically a 20K ohm input impedance.

In the circuit illustrated in Fig. 5-3, a 40 dB gain is obtained. A split power supply is necessary and should be capable of but not exceed plus and minus 24 volts A good heatsink is needed to keep the 540 operating within its safe temperature range. Since the IC has a wide power bandwidth, the layout should best be done on a PC board to avoid oscillation. In Appendix 2 you will see the table giving you the pin connections for this and other ICs mentioned in this book.

A socket is recommended for the IC, as you may want to use this IC to build another amplifier, which is described later in this chapter. In that case, the amp delivers 35 watts per channel by using a pair of series transistors.

15 TO 25 WATT DARLINGTON AMPLIFIERS

With the schematic devised by Motorola, you can choose varied amounts of power you want from basically the same circuit. Only a few components are changed—actually a total of six. Refer to Fig. 5-4.

In one of the previous circuits in this chapter, a pair of transistors were connected as a Darlington pair, by interconnecting the two collectors. In this circuit, the two transistors are in the same case and are connected internally.

Use Table 5-1A to determine the values of the components you are going to use and therefore the output power of the amplifier. Table 5-1B tells you what to expect from each of the three amplifiers. As usual, the greater output must be paid for. In this case it means that the higher the output, the greater must be the input. So, before you jump into building the 25-watter, you must know the maximum capability of your preamp.

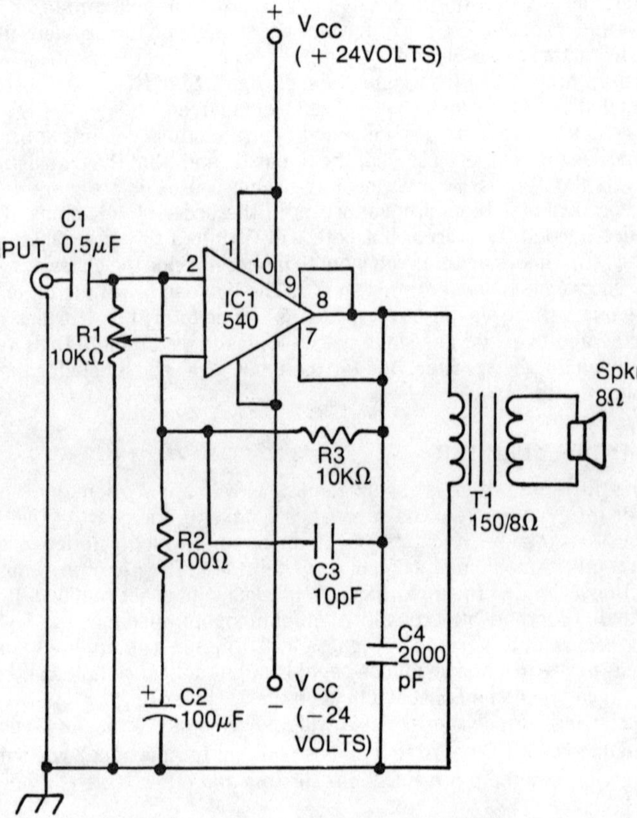

PARTS LIST

C1—0.5 μF mylar tubular capacitor

C2—100 μF, 35-volt electrolytic capacitor

C3—10 pF silvered mica capacitor

C4—2000 pF silvered mica capacitor

IC1—Operational amplifier (Signetics 540, or equal)

R1—10,000 ohm audio taper potentiometer

R2—100 ohm, 1/2 watt carbon resistor

R3—10,000 ohm, 1/2 watt carbon resistor

SPKR—8-ohm speaker

T1—150/8 ohm interstage transformer

Fig. 5-3. A 1-watt IC amplifier requiring a double-ended power supply.

Description

This is simply a power amplifier and all of the audio controls such as bass, treble, balance, as well as the volume must be on the preamp. Seven

Table 5-1. Component Changes Required For the Various Output Levels Are Shown Gives the Run Down of the Performance Characteristics of All Three Amplifiers.

Component	15 W	20 W	25 W
V_{CC}	42 V	46 V	48 V
R2	220 k	160 k	160 k
R5	120 Ω	180 Ω	180 Ω
R13	180 Ω	220 Ω	220 Ω
R16	180 Ω	220 Ω	220 Ω
Q6	MJE800	MJE1100	MJE1100
Q7	MJE700	MJE1090	MJE1090
*Heatsink	6.5°C/W	4.8°C/W	3.6°C/W
R8	Adjusted for 20-30 mA Collector Current for Q6		

A

Amplifier Performance Characteristics

Reference Figure 5-4	15 W 42 Vdc	20 W 46 Vdc	25 W 48 Vdc
1. Idle Current (nominal no-signal)	20 mA	20 mA	20 mA
2. Current Drain at Rated Power	640 mA	760 mA	800 mA
3. Typical Input Impedance	50 k	56 k	56 k
4. THD at Rated Output Power 20 Hz or 1 kHz to 20 kHz	$<$1%	$<$1%	$<$1%
5. IM Distortion at 60 and 7000 Hz 4:1 ratio at Rated Power	$<$1%	$<$1%	$<$1%
6. -3 dB Bandwidth	23 Hz-170 kHz	27 Hz-180 kHz	28 Hz-190 kHz
7. Typical input sensitivity for rated output power	200 mV	220 mV	230 mV
8. Maximum output power at 5% THD without current limiting	21.2 Watts	28.5 Watts	32.4 Watts
9. Maximum output power at 5% THD with current limiting	21 Watts	28 Watts	30 Watts
10. Power Supply Ripple Rejection	29 dB	34 dB	33 dB
11. Short Circuit Power Supply Current with Current Limiting	1.5 Amps	1.5 Amps	1.5 Amps

B

transistors are used. Q1 is a silicon RF transistor. PNP type. Q2 can be any general purpose silicon audio amplifier. Note that this one and Q4 are NPN. For Q2 choose one that has low noise. if you can. Any noise here will be amplified in the succeeding stages. Q3 is an NPN Darlington amplifier. Q5 is a general purpose PNP. silicon. low frequency RF and audio amplifier. The last two. Q6 and Q7. are a pair of complementary power transistors. also Darlingtons. Using these Darlingtons reduces the total number of transistors. presenting high gain with matched internal characteristics.

A few precautions should be observed: although the amplifier is short-circuit protected. it is best to place a substitute load on the output of the amp before turning on the power. An 8 or 15-ohm resistor rated at or above the highest wattage you are apt to be getting makes a good load. It is best not to use your good speakers as the load during testing. since there is always the

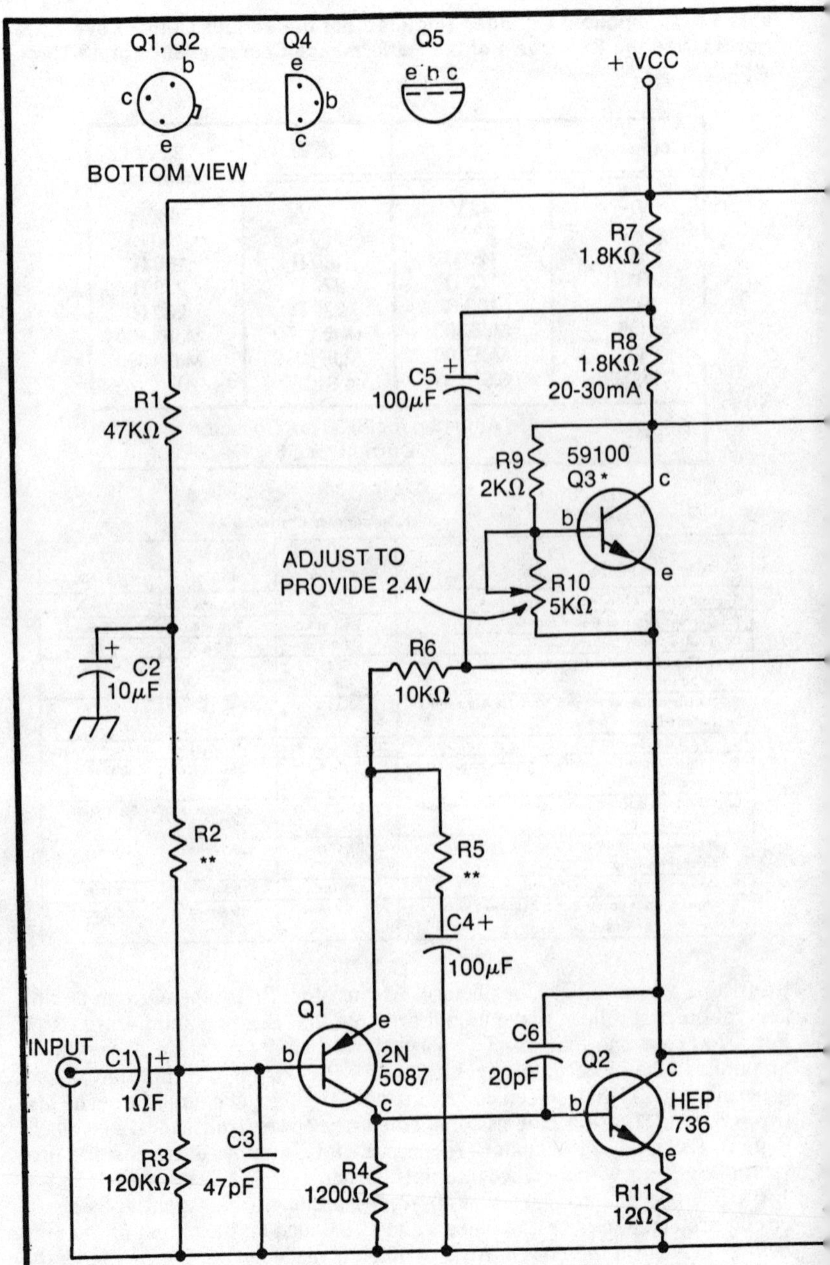
Q1, Q2
b
c
e
BOTTOM VIEW
Q4
e
b
c
Q5
e b c
+ VCC
R7
1.8KΩ
R8
1.8KΩ
20-30mA
C5
100µF
R1
47KΩ
R9
2KΩ
59100
Q3 *
c
b
e
ADJUST TO
PROVIDE 2.4V
R10
5KΩ
C2
10µF
R6
10KΩ
R2
**
R5
**
C4+
100µF
Q1
e
b
2N
5087
c
INPUT
C1
1ΩF
C6
20pF
Q2
c
b
HEP
736
e
R3
120KΩ
C3
47pF
R4
1200Ω
R11
12Ω

Fig. 5-4. Circuit for an amplifier using Darlington transistors. Choice of components permits building either a 15, 20, or 25 watt amplifier.

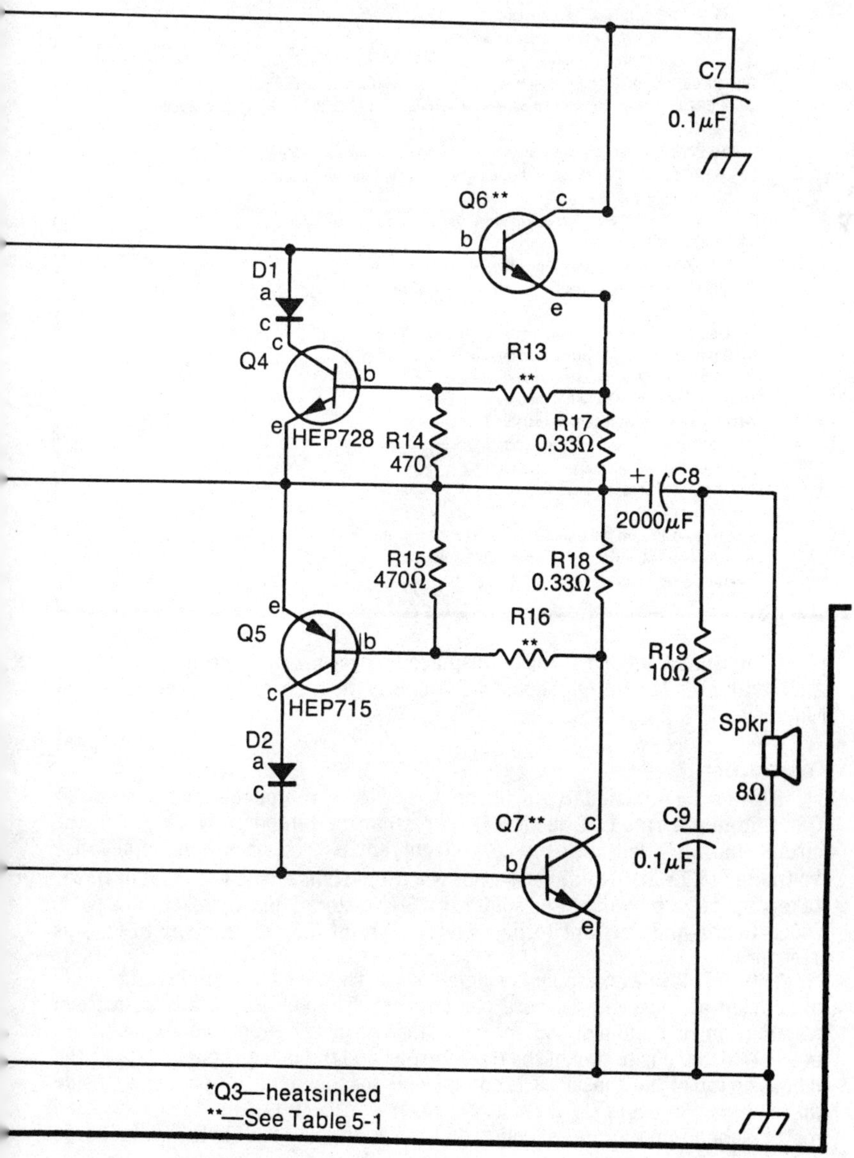

PARTS LIST

C1—1 μF, 75-volt electrolytic capacitor
C2—10 μF, 75-volt electrolytic capacitor
C3—47 pF silvered mica capacitor
C4, C5—100 μF, 75-volt electrolytic capacitor
C6—20 pF silvered mica capacitor
C7, C9—0.1 μF mylar tubular capacitor
C8—2000 μF, 75-volt electolytic capacitor
D1, D2—50-volt (PIV), 1 ampere diode (Radio Shack 276-1101, or equal)
Q1—2N5087 low-noise, high-gain transistor (Radio Shack 276-2022, or equal)
Q2—HEP736 general purpose amplifier silicon transistor (Radio Shack 276-2009, or equal)
Q3—S9100
Q4—HEP728 general purpose amplifier transistor (Radio Shack 276-2016, or equal)
Q5—HEP715 general purpose audio amplifier transistor (Radio Shack 276-2034, or equa!)
Q6, Q7—See Fig. 5-6
R1—47,000 ohm, 1/2 watt carbon resistor, 10% tolerance
R2—See Fig. 5-6
R3—120,000 ohm, 1/2 watt carbon resistor, 10% tolerance
R4—1200 ohm 1/2 watt carbon resistor, 10π tolerance
R5—See Fig. 5-6
R6—10,000 ohm, 1/2 watt carbon resistor, 10% tolerance
R7, R8—1800 ohm, 1/2 watt carbon resistor, 10% tolerance
R9—200 ohm, 1/2 watt resistor, 10π tolerance
R10—5000 ohm miniature potentiometer
R11—12 ohm, 1/2 watt carbon resistor, 10% tolerance
R12—390 ohm, 1/2 watt carbon resistor, 10% tolerance
R13—See Fig. 5-6
R14, R15—470 ohm, 1/2 watt carbon resistor, 10% tolerance
R16—See Fig. 5-6
R17, R18—0.33 ohm, 1/2 watt carbon resistor, 10% tolerance
R19—10 ohm, 1/2 watt carbon resistor, 10% tolerance
SPKR—8 ohm speaker

possibility that a sudden surge of misplaced power may rip the speaker cones. Another hint: Q2 and Q3 should be properly heat-sinked to avoid thermal runaway.

Construction

R8 must be adjusted to provide between 20 to 30 mA of collector current for Q6. A mini-pot rated at about 2.5K ohms can be temporarily placed in the circuit and adjusted until the correct current is read on a high impedance voltmeter (VTVM). Remove the pot when the right amount of current flows, take a resistance reading and solder in a 5% reistor of that value. Using a pot is much faster and less apt to damage the circuit than by clipping in various resistors.

The 48-volt power supply is single-ended. The transformer should be able to provide at least one ampere per channel. To avoid hum, use a shielded transformer or build a shield around it and the power supply components. An even better way is to mount the transformer and the large capacitors with the diodes on top of the chassis. The components for the amplifier proper go inside the chassis. To dress up the case, a cover can be fitted made of expanded metal or bend an aluminum cover. In the latter case, drill holes in the cover above the transformer to allow the heat generated by the transformer to dissipate. Keep the power line away from the audio inputs.

The components can all fit on a 3.5 × 4-inch PC board. See Fig. 5-5. For stereo, merely add another board. The same power supply will handle both boards. The power supply components will take another 5 × 5 inches. Allow room for the interconnections. For the mono version, the chassis needs to measure only 5 × 7 × 3 inches. For stereo 5 × 10 × 3 inches will do nicely.

No controls are located on the amplifier, but you will need a SPST switch to turn on the power. You might want to add a power outlet on the back of the amplifier so you can plug in the preamp. If your preamp uses the same voltage, i.e., 48 volts, you can use the same power supply—provided your transformer is able to put out the necessary power. Should the amplifier's output voltage be higher than needed for the preamp, a current regulator can be used or else calculate the proper resistor and zener diode combination as explained in Chapter 4.

If the preamp's voltage is higher than the amplifier's then reverse the procedure: feed the amp from the preamp's power supply using a regulator or zener and resistor combination.

To make certain you will have the necessary power, use a current meter to measure what the preamp and amp draw. Multiply this current by the voltage to obtain the power (in watts) drawn. Always add some 20% to avoid surges that may blow diodes or overheat a transformer.

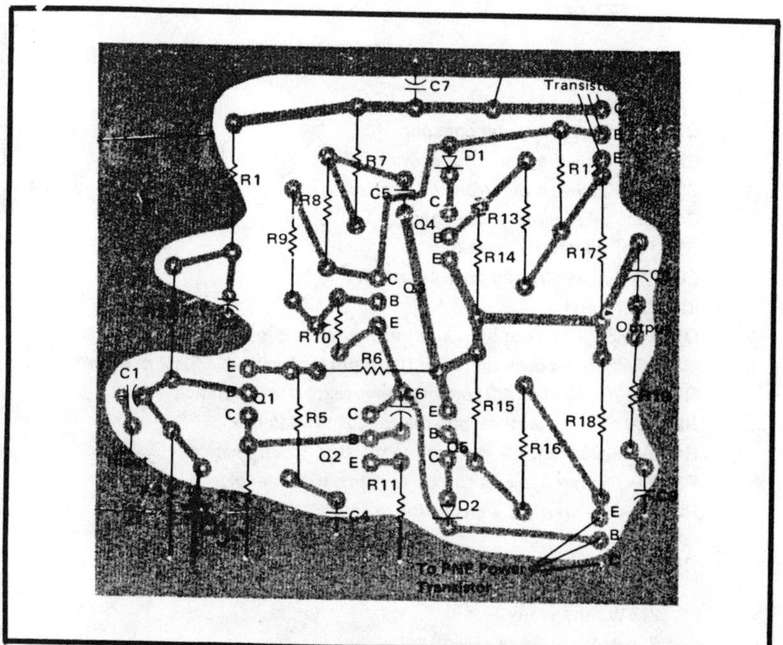

Fig. 5-5. Printed circuit pattern for the Darlington amplifier. The same pattern can be used for all three power levels.

35WATT IC AMPLIFIER

A somewhat different approach to a solid-state amplifier is the one in the schematic of Fig. 5-6 Here the IC, a Signetics 540, which we used in the one-watt amp, is brought into play again. In this case, it is employed to drive a pair of complementary output transistors. The 540 is a monolithic class AB

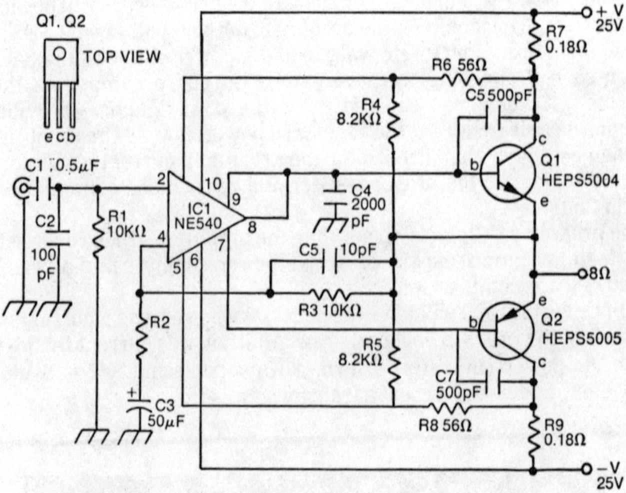

PARTS LIST

C1—0.5 μF mylar tubular capacitor

C2—100 pF silvered mica capacitor

C3—50 μF, 35-volt, electrolytic capacitor

C4—2000 pF silvered mica capacitor

C5—10 pF silvered mica capacitor

C6, C7—500 pF silvered mica capacitor

IC1—SE/NE540

Q1—HEPS5004 power amplifier transistor (Motorola)

Q2—HEPS5005 power amplifier transistor (Lafayette 32-05267, or equal)

R1, R3—10,000 ohm, 1/2 watt carbon resistor, 10% tolerance

R2—100 ohm, 1/2 watt carbon resistor, 10% tolerance

R4, R5—8200 ohm, 1/2 watt carbon resistor, 10% tolerance

R6, R8—56 ohm, 1/2 watt carbon resistor, 10% tolerance

R7, R9—0.18 ohm, 1/2 watt carbon resistor, 10% tolerance

A parts kit based on this design is available from:

Southwest Technical Products
219 W. Rhapsody
San Antonio, Texas 78216

Fig. 5-6. Schematic for a 35-watt amplifier using one IC per channel.

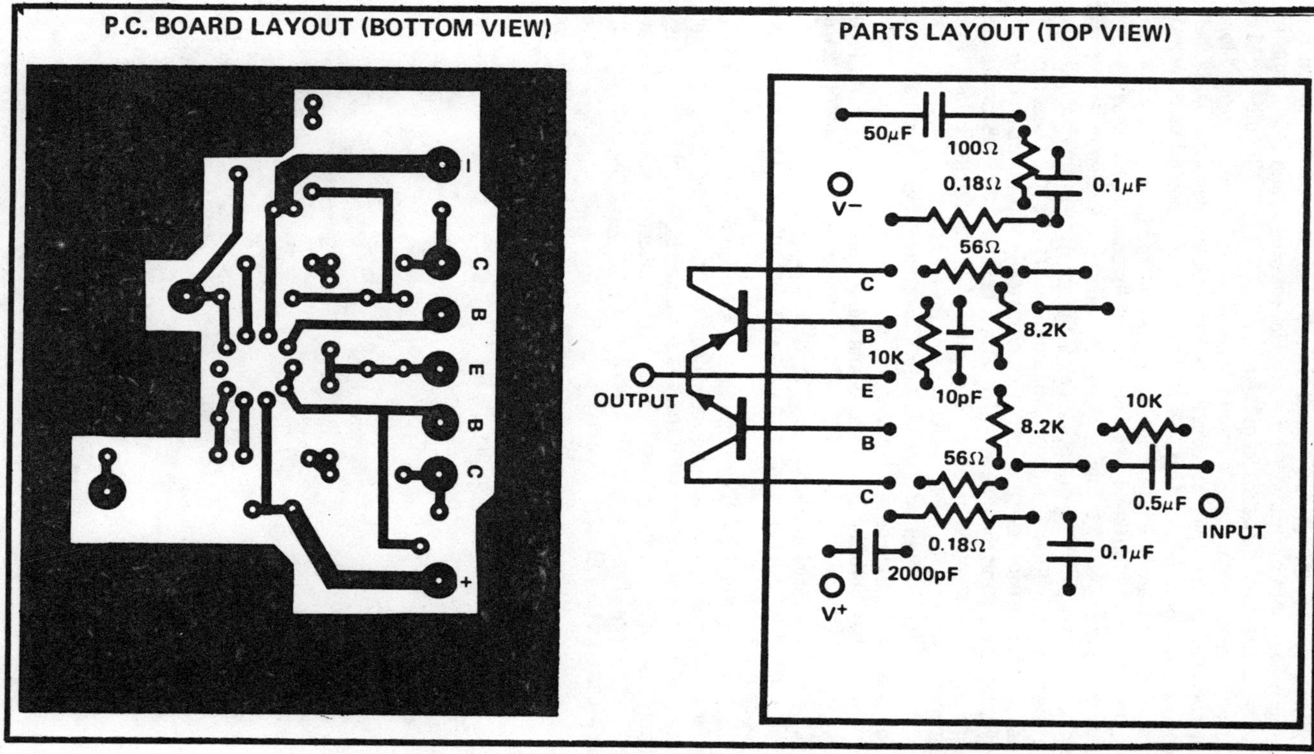

Fig. 5-7. PC pattern and parts layout for the 35-watt amplifier.

amplifier, with a low standby current, retaining a high output current capability with internal current limiting. The schematic and PC layout are furnished courtesy of Signetics from their book, Linear ICs, Volume One. The PC board layout shown is for only one channel. So, two boards are needed for a stereo version. The component count is low and no exotic parts are demanded. A double ended power supply capable of putting out plus and minus 25 volts is required. Only 0.5 ampere is needed for both channels.

The 540 is an 8-pin IC and a socket in which to mount it is necessary, or at least, recommended. Q1 and Q2 are a pair of complementary power transistors. They should be mounted against the chassis with silicone grease to provide heat dissipation. The chassis in this case acts as a giant heat sink. The cases of the transistors are not connected electrically to any pin, so no insulation is required. Bend the pins away from their case with care, using a pair of long nose pliers, so as not to break them off. Soldering should be done with a heat sink attached to the lead to avoid heating the transistor.

The resistors should be of at least 10% tolerance; 5%, if you can obtain them reasonably would be much better.

Fig 5-7 has both the PC board layout (exact size) and the parts layout. These figures are courtesy of Signetics also. Use a low wattage soldering iron and solder the transistors and the IC connections last.

The amplifier is meant to be used with an 8-ohm speaker. Do not connect it to a 4-ohm speaker.

Before inserting each board into the chassis, run a test on it. It is much easier to do it now than after it has been connected to the front and rear panels. It also provides a check in that if there is some operational difficulty—this is engineerese for...."it doesn't work"—after it is connected to the case the fault lies inside there. The unit will work very well with lower voltages, although you will not get the full output power.

CHAPTER 6
PREAMPLIFIERS

As the name suggests, a preamplifier presents a stage, or sometimes two stages, of amplification before the amplifier. In many commercial units, the amp and preamp are combined in the same chassis. There is no reason for you not to do the same.

The reason for the existence of the preamp is that many signal sources, such as a magnetic phonograph cartridge, a microphone, or a tape head put out a very low level signal. In fact, it is so low that the amplifier, if this signal were to be connected directly to it, could not even "see" it and so could not amplify it. So, as Fig. 6-1 shows, the preamp is connected between the low level signal source and the input of the amplifier.

For the two units to work together, the preamp must put out a signal of sufficient power that the amplifier can amplify it. Shielded cable must be used between the signal source to the preamp and from the preamp to the amplifier. To avoid high frequency losses, make sure your units are close together. They should not be more than two feet apart.

MAGNETIC PHONO PREAMP

If your present amplifier has only high level phono inputs, its use is limited to a ceramic cartridge. You have a magnetic cartridge, so what do you do? Build yourself the phono preamp shown in Fig. 6-2. Magnetic cartridges are vastly superior to ceramics. Their output, however, is much lower than the ceramics'. The latter have outputs ranging from 100 mV to as high as two volts. They require no preamplification. Magnetics produce only from 3.5 mV to no higher than 8 mV. This means that ceramics are from 30 to 250 times more powerful.

To make up for this inadequacy, the circuit in Fig. 6-2 offers both gain and equalization. The heart of the circuit is the LM381. Only one of these ICs is employed, since there are two separate preamps in the package. A PC board is

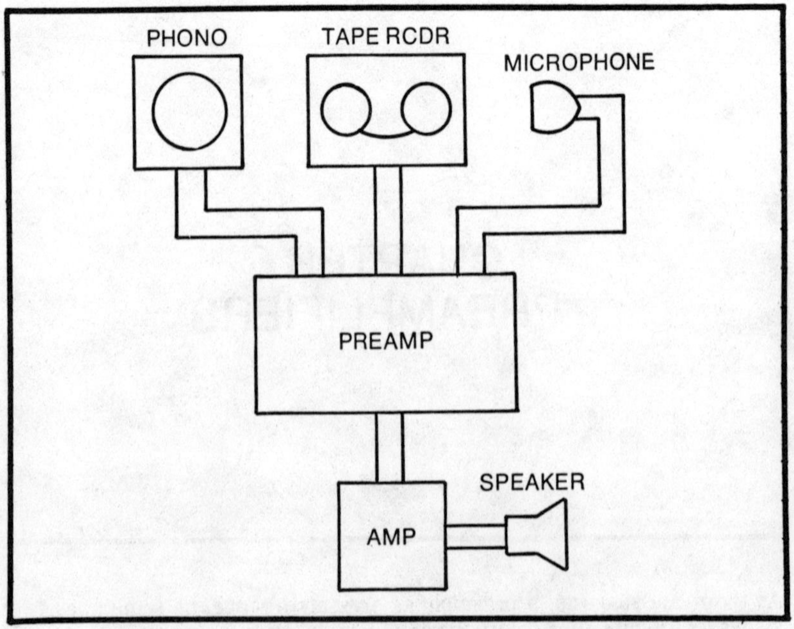

Fig. 6-1. Block diagram showing a preamp in a high-fidelity system.

advised, to prevent oscillation and hum. This is a common trouble with a high gain circuit.

We mentioned the gain offered by the LM381, but also briefly noted was the fact that this preamp equalizes the output from the magnetic cartridge. Equalization is needed to obtain the desired overall frequency response. This equalization circuit is designed to compensate for the preemphasis introduced at the recording studio. The equalization in our circuit is determined by resistors R4, R5, R6 and capacitors C3 and C4. The recording curve is different from that used in a broadcasting studio, which is why the equalization circuit for the microphone preamp contains a different combination of resistors and capacitors. This results in a frequency curve which is the analog of the recording curve.

The component count is low. Only five resistors and four capacitors are needed for each channel. A PC board 3 × 3 inches is large enough for the circuit—with the exception of the power supply. To avoid hum pickup it is best to shield the power supply or keep it at a distance from the other components. A single-ended 24-volt supply is sufficient. The schematic in Fig. 6-3 can power this preamp very well.

The LM381 has two stages of amplification built within it, so this will give you adequate gain for your magnetic cartridge.

Each half of the preamp should have its own output and ground. Do not use a common ground. Use short leads and shielded cable from the input to the PC board. A 10 × 5 × 3 inches high chassis is very roomy. It allows the power supply elements to be put to one side with the board at the other.

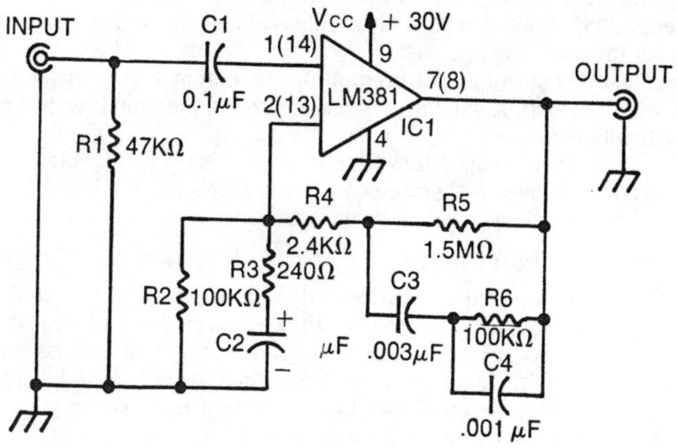

PARTS LIST
C1—0.1 µF mylar tubular capacitor
C2—20 µF, 50-volt electrolytic capacitor
C3—0.003 µF mylar tubular capacitor
C4—0.001 µF mylar tubular capacitor
IC1—LM381 dual preamp (Motorola HEP C6080P, or equal)
R1—47,000 ohm, 1/4 watt carbon resistor, 20% tolerance
R2, R6—100,000 ohm, 1/4 watt carbon resistor, 20% tolerance
R3—240 ohm, 1/4 watt carbon resistor, 20% tolerance
R4—2400 ohm, 1/4 watt carbon resistor, 20% tolerance
R5—1.2 megohm, 1/4 watt carbon resistor, 20% tolerance

Fig. 6-2. Circuit for a magnetic cartridge preamp. Only one IC is used in this circuit.

After a signal has been amplified a stage or two the result is a "flat" output. There are many purists who feel this is the way it should be. No additional coloring or frequency controls should be added. is their cry. In theory this is fine. But. our listening area adds its own coloring to the sound emanating from the speakers. Overstuffed furniture can absorb high frequencies to an unbelievable degree. The shape of a room also does things to the higher frequencies. To compensate for these imperfect conditions you might want to exercise some control over this "flat" output.

For this purpose. the circuit in Fig. 6-4 does exactly that. It has both bass and treble controls. each offering boost and cut. The treble control is an analog of the bass. Audio taper potentiometers should be employed for these controls. They should be low noise and be sprayed with some contact cleaner. if they exhibit the least tendency to sound like sand underfoot every time the controls are rotated.

A balance control is provided to adjust the volume so the individual speakers sound at the same level regardless of how much farther one might be from your favorite listening post.

The volume control on the preamplifier is optional. It offers the potential of reducing the output should it be too much for the amplifier to handle this level without distortion.

The resistors and capacitors which form the bass and treble controls are mounted right on the pots themselves. See the photograph in Fig. 6-5. All of this fits right in the original chassis with no squeezing.

VARIABLE SCRATCH FILTER

There are lots of old 78 rpm records still around: many are collectors items. They usually have one trouble: they are scratchy and almost unplayable on modern playback equipment. Then too, you may own some favorite records that have been played very much and they too are scratchy. The circuit in Fig. 6-6 has been designed specifically for that purpose, so you can enjoy those old records. The circuit will not restore the records to the pristine quality they had, but it will remove the worst noise. The filter is variable so you can put in just as much scratch control as you feel that record needs. The trick is not to filter out more than necessary, as you are also filtering out music that lies in that frequency range.

For the 78s you need only one audio channel, but for a beat-up stereo disc you will need two of these circuits. It is so simple and requires so few components that it is hardly worthwhile to make a PC board. Layout is not important

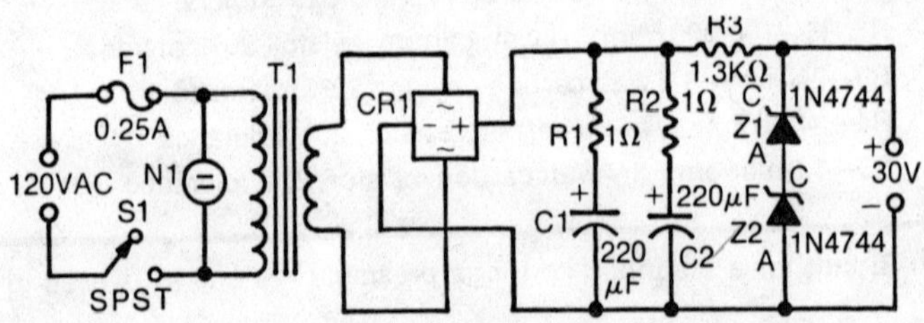

PARTS LIST

C1, C2—220 µF, 50-volt electrolytic capacitor

CR1—100-volt (PIV), 0.5 ampere bridge rectifier

F1—0.25 ampere fuse

N1—Neon pilot lamp (Radio Shack 272-703, or equal)

R1, R2—1 ohm, 1 watt carbon resistor, 20% tolerance

R3—1300 ohm, 1 watt carbon resistor, 20% tolerance

T1—24-volt, 0.25 ampere transformer

Z1, Z2—1N4744, 15-volt, 1 watt zener (Radio Shack 276-564, or equal)

Fig. 6-3. Thirty-volt power supply for the phono preamp.

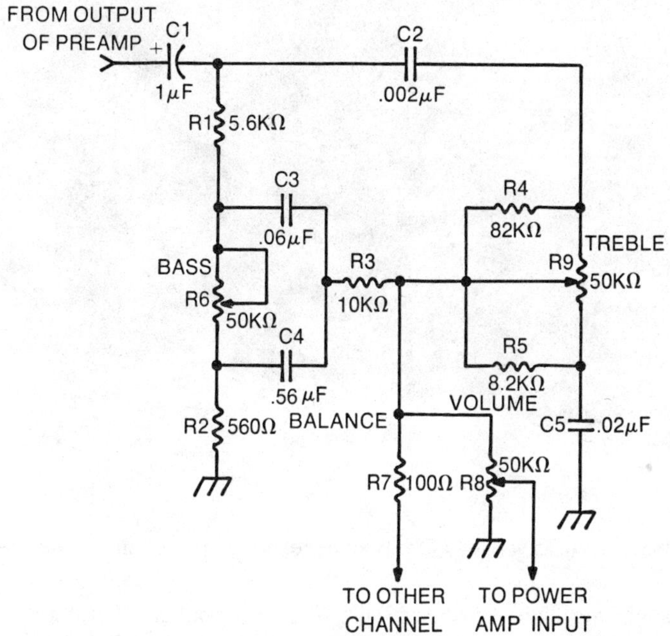

PARTS LIST

C1—1 μF, 50-volt electrolytic capacitor
C2—0.002 μF mylar tubular capacitor
C3—0.06 μF mylar tubular capacitor
C4—0.56 μF mylar tubular capacitor
C5—0.02 μF mylar tubular capacitor
R1—5600 ohm, 1/4 watt carbon resistor, 20% tolerance
R2—560 ohm, 1/4 watt carbon resistor, 20% tolerance
R3—10,000 ohm, 1/4 watt carbon resistor, 20% tolerance
R4—82,000 ohm, 1/4 watt carbon resistor, 20% tolerance
R5—8200 ohm, 1/4 watt carbon resistor, 20% tolerance
R6, R8, R9—50,000 ohm, audio taper potentiometer
R7—100 ohm, 1/4 watt carbon resistor, 20% tolerance

Fig. 6-4. Typical bass and treble controls for magnetic phono.

and the unit can be placed anywhere near the record player. The only requirement is that you can get at control potentiometer R2 and switch S1 easily. The circuit as it stands provides 18 dB per octave of filtering with the variable control giving you cut-off frequencies ranging from 6 to 15 kHz. The filter should

Fig. 6-5. Wired version of the phono preamp with tone controls included.

feed into the high impedance input of your amplifier. If you are using a magnetic cartridge you will need at least one stage of preamplification before the amplifier's inputs.

You will notice that the circuit is entirely a passive one; not only does it not amplify the signal, a loss will occur due to the amount of resistance built into the circuit.

Input and output jacks should be selected that will mechanically match your existing equipment. A plastic box, which you might find around the house, will do perfectly well as a housing for the scratch filter.

ADD-ON FILTER CONTROLS

This unit is designed to be employed as an "outboard" piece of equipment to remove scratches and rumble in sophisticated set-ups. The unit accomplishes this by filtering out the frequencies in the range in which the noise makers exist. Studying Fig. 6-7 shows that S1 cuts IC1 into the circuit with R5, R6, R7, C1, C2 and C3. This circuit, or rather this section of the circuit, removes the sound of scratches that appear on an old record or one that has been abused. S2 does the same with IC2 and its attendant components, C4, C5, C6, R8, R10 and R11. The switch, by throwing these components into the circuit, removes the rumble. Some record players have a tendency to rumble, but with S2 in the ON position the rumble disappears.

C1 and C2 have no effect on mid-frequencies. The resulting gain is unity, due to the high input impedance of the IC with its low output impedance. Feedback occurs at higher input frequencies, with C1 and C2 bypassing the signal to ground. Feedback goes from the output of IC1 through C3 back to the input at pin 3.

In the rumble section of the filter, the capacitors C4, C5, C6 offer a low impedance at mid-frequencies with a unity gain. The capacitors increase in impedance and the output is lowered as the input frequency decreases.

In the case of both ICs, the output at pin 6 returns the output directly to the inverting input at pin 2. This results in a non-inverting buffer with a high input impedance and a low output impedance.

Q1 has two functions: it forms a emitter-follower driving the scratch filter section: it presents the necessary DC bias for IC1. Depending on the position of S2, the rumble filter is driven either by the scratch filter section or, if S1 is in the OFF position, it is driven by Q1. R8 and R11 in the rumble filter section furnish the DC for the circuit. C1 is a decoupling element.

Construction

The circuit shown is for one channel. If a stereo version is planned, the unit should be wired preferably on two boards, either PC or perforated. It is much easier to tackle a project in small bites...which is why two boards are recom-

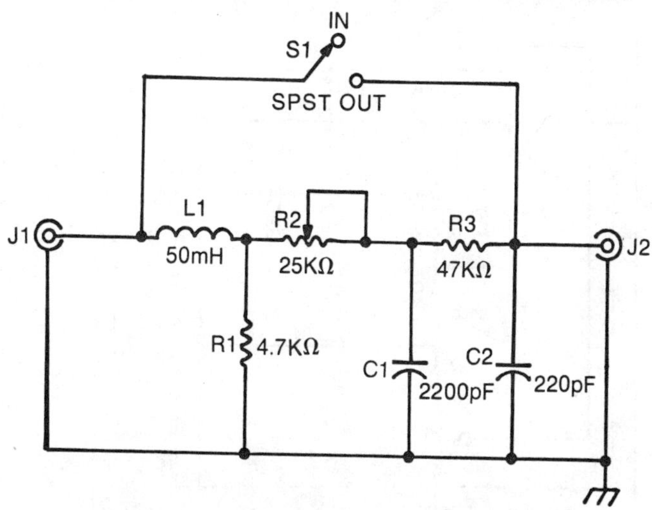

PARTS LIST

C1—2200 pF silvered mica capacitor

C2—220 pF silvered mica capacitor

J1, J2—Jack, to match existing connector type

L1—50 mH inductor

R1—4700 ohm, 1/2 watt carbon resistor, 20% tolerance

R2—25,000 ohm, audio taper potentiometer

R3—47,000 ohm, 1/2 watt carbon resistor, 20% tolerance

S1—Single-pole, single-throw switch

Fig. 6-6. Schematic for a filter designed to eliminate scratches from an old disc.

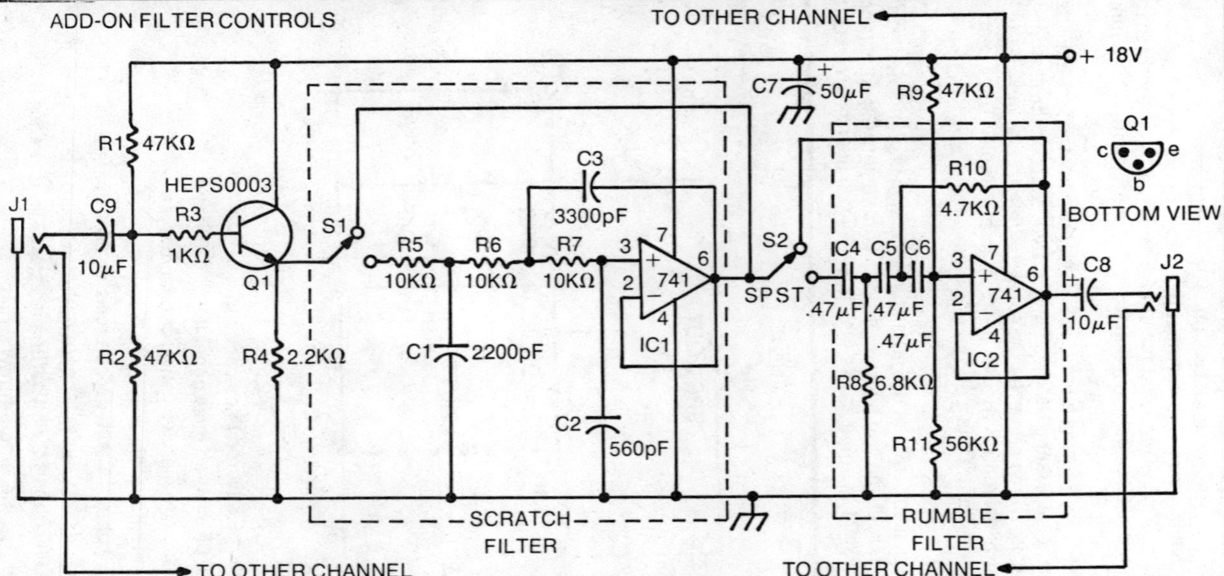

PARTS LIST

C1—2200 pF silvered mica capacitor
C2—560 pF silvered mica capacitor
C3—3300 pF silvered mica capacitor
C4, C5, C6—0.47 μF mylar tubular capacitor
C7—50 μF, 25-volt electrolytic capacitor
IC1, IC2—741 operational amplifier (Radio Shack 276-010, or equal)
J1, J2—Stereo jacks
Q1—HEPS0003 audio amplifier (Radio Shack 276-2014, or equal)
R1, R2, R9—47,000 ohm, 1/4 watt carbon resistor
R3—1000 ohm, 1/4 watt carbon resistor
R4—2200 ohm, 1/4 watt carbon resistor
R5, R6, R7—10,000 ohm, 1/4 watt carbon resistor
R8—6800 ohm, 1/4 watt carbon resistor
R10—4700 ohm, 1/4 watt carbon resistor
R11—56,000 ohm, 1/4 watt carbon resistor
All resistors 20% tolerance
S1, S2—Single-pole, double-throw switch

Fig. 6-7. An all IC scratch and rumble filter circuit to add to output of phono preamp. It can also be made part of the power amplifier.

mended. The board should not be laid out until all the components have been purchased. For one reason, the 741 is available in two versions, the DIP (Dual In-line Package) or in the more usual round type (the TO-5 case).

CERAMIC PICKUP AMPLIFIER

The amplifier illustrated in Fig. 6-8 is suitable for a child's phonograph, which uses a ceramic or crystal pickup. No preamp is necessary due to the high output of these cartridges. Instead, the signal goes directly to the amplifier, an IC with good noise characteristics. This signal is in turn fed directly to an 8-ohm speaker. The gain is internally fixed at 34 dB, which is adequate for the purpose. The volume will not be ear shattering, but is that bad?

In spite of its small size and simplicity of design, the amplifier boasts of a volume control and a simple tone control. Apart from these two pots, three external components complete the unit. The whole thing, with its two 9-volt batteries, can be tucked inside the phonograph case. Quiescent supply current is only 7 mA at 18 volts, so the batteries will be long lived.

The IC, an LM380 op amp, is short circuit proof with internal thermal limiting.

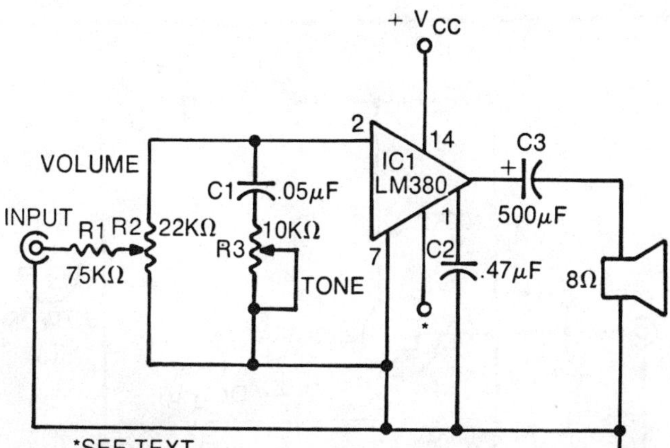

PARTS LIST

C1—0.05 μF mylar tubular capacitor

C2—0.47 μF mylar tubular capacitor

C3—500 μF, 35-volt electrolytic capacitor

IC1—LM380 audio power amplifier (Lafayette 32 P 06588V, or equal)

R1—75,000 ohm, 1/4 watt, low noise metal film resistor

R2—22,000 ohm, audio taper potentiometer

R3—10,000 ohm, linera taper potentiometer

Fig. 6-8. Preamp for ceramic pick-up. Suitable for a child's phonograph.

Three pins on each side of the DIP should be heat sinked by connecting them to the PC board, which has a minimum of six square inches of copper foil. These pins are 3, 4, 5, 10, 11, and 12.

IMPEDANCE MATCHING CIRCUIT

Feeding a high impedance microphone into a recorder or public address system with a long length of cable results in the loss of the high frequency tones. This is caused by the capacitance of the cable. The obvious way to cure this is to have the mike close to its input. This is not always possible. Or you could use a low impedance mike, but you may not want to go to the expense for just that special purpose. The same problem exists if you want to use a tape recorder with a high impedance output and you want to connect it to a low impedance amplifier. A severe mismatch will mean a great loss of power.

The circuit illustrated in Fig. 6-9 will solve each of the problems listed above. It is simple to build and simple in its operation.

The circuit is an emitter-follower with no inversion of the signal. Input impedance is about 50K for load inputs whose impedance is 5K or a little over. This is about right for most amplifiers.

Only three resistors and two capacitors make up the circuit. The only transistor is a 2N3638A, although almost any PNP silicon will do just as well.

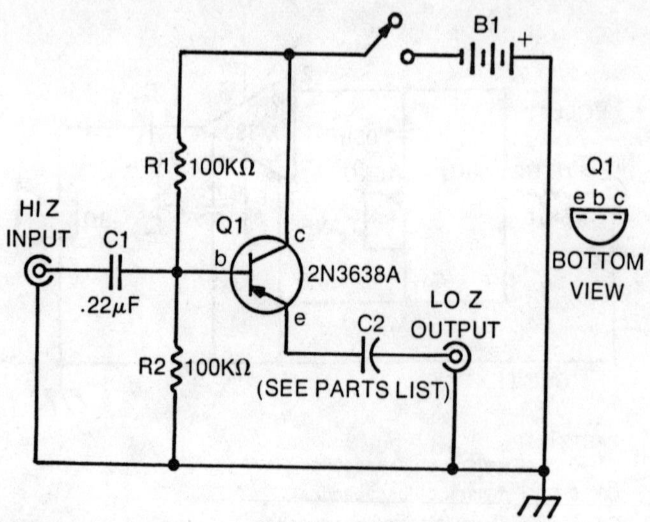

PART LIST

B1—9-volt transistor radio battery

C1—0.22 μF mylar tubular capacitor

C2—Any value from 1 to 2.2 μF, 25-volt electrolytic capacitor

Q1—2N3638A general purpose silicon transistor (Radio Shack 276-2034, or equal)

Fig. 6-9. Circuit to match a high-impedance output with a low impedance input.

With this circuit we get some amplification and a change of impedance, but you have to pay for it. In this case the cost is a voltage loss. This means the applied signal must be of such a level that it can overlook the loss. The unit should be encased in a metal box with connectors for input and output that will mechanically match to whatever you are going to connect this impedance matchmaker.

PREAMP FOR PHONO, FM AND TAPE INPUTS

This is a high quality unit incorporating all the controls needed if you are using a driver amplifier which has none. The preamp has, besides treble and bass controls, a scratch filter for noisy records, plus a volume control circuit which incorporates a loudness switch. See Fig. 6-10. When switched in, the lower tones are enhanced at low output levels. You may have noticed with most amplifiers that when the volume is turned down the treble remains dominant while the bass seems to have disappeared. This switch corrects this effect. With the switch in the ON position, the volume control attenuates all tones equally. All the controls are independent of each other so no interaction exists.

The phono input is designed for a magnetic pickup supplying an input signal of 5 mV or over. This is average for most magnetic cartridges. A 5 mV input delivers a 1-volt output, more than enough to drive most amplifiers.

Provision is made for an FM input and for tape. The tape recorder can be switched from IN to OUT. This permits recording through the preamp or playing back as well. The input in this case must provide at least 300 mV to produce the 1-volt output from the preamp.

Circuit Description

The first section consists of Q1 and Q2 in a two-stage direct-coupled input circuit. This section is used only by the magnetic pickup and provides not only the extra amplification to drive such a low level signal, but frequency compensation as well. This is accomplished by the frequency shaping network in the feedback circuit of Q2.

R10 is a level control in the output circuit of the same transistor. This feeds pot R12, a loudness control, through the selector switch, S1. The switch eliminates the first two stages of the preamp.

The loudness control drives the tone control circuits. Tape recordings can be made without affecting loudness or volume settings.

Treble and bass controls offer 10 dB of boost and 15 dB of cut for the bass and treble frequencies, respectively. You will note that one control has no effect on the other. A flat response is obtained by placing the controls in the middle position. Q3 and Q4 operate a collector-to-base network which, with the tone controls, provides an excellent overall tonal response.

Construction

This is not a simple project and is not recommended for the beginner. However, it is worth the effort to build this unit, due to the excellent results obtained. To the partially experienced hobbyist, we suggest you use the "bite-size" approach. It makes the total circuit much simpler to build. As we explained, Q1 and Q2 form one section with their external components; Q3 and Q4 another section. The input section, with the loudness and volume controls and its associated components, make up the third section, and a very simple one at that. The fourth section is made of the tone controls. You will see that these controls are quite similar to the outboard controls presented in another

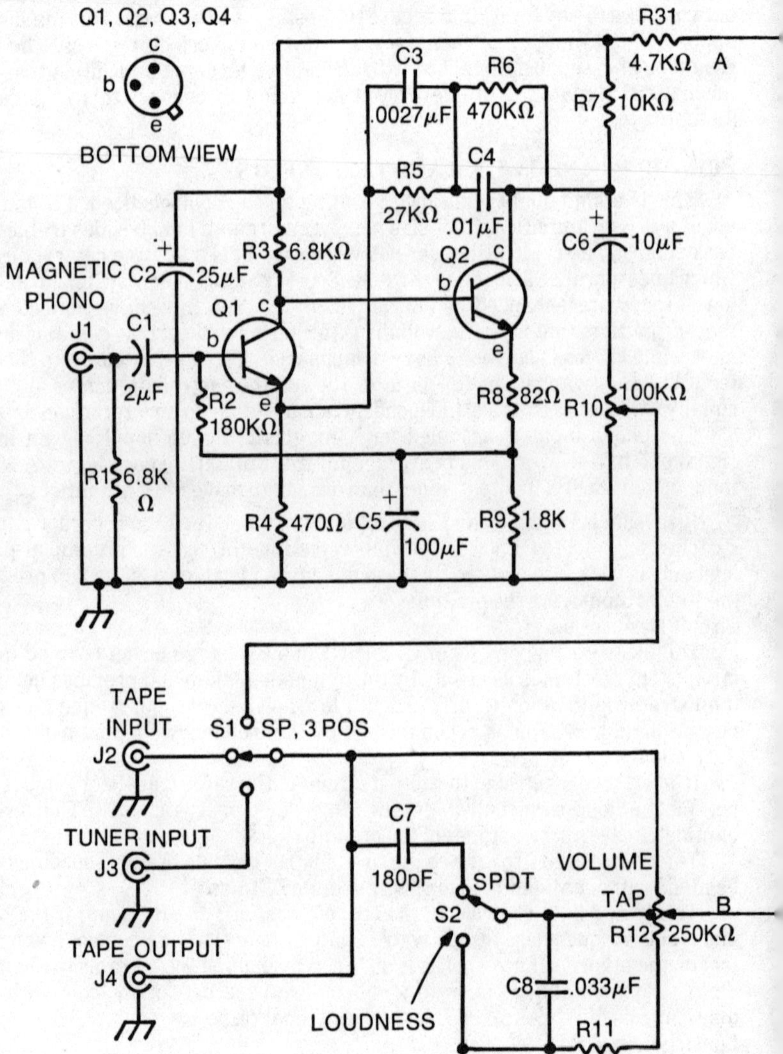

Fig. 6-10. A quality preamp suitable for tuner, tape in, tape out, and phono. Circuit incorporates bass, treble, and scratch filters.

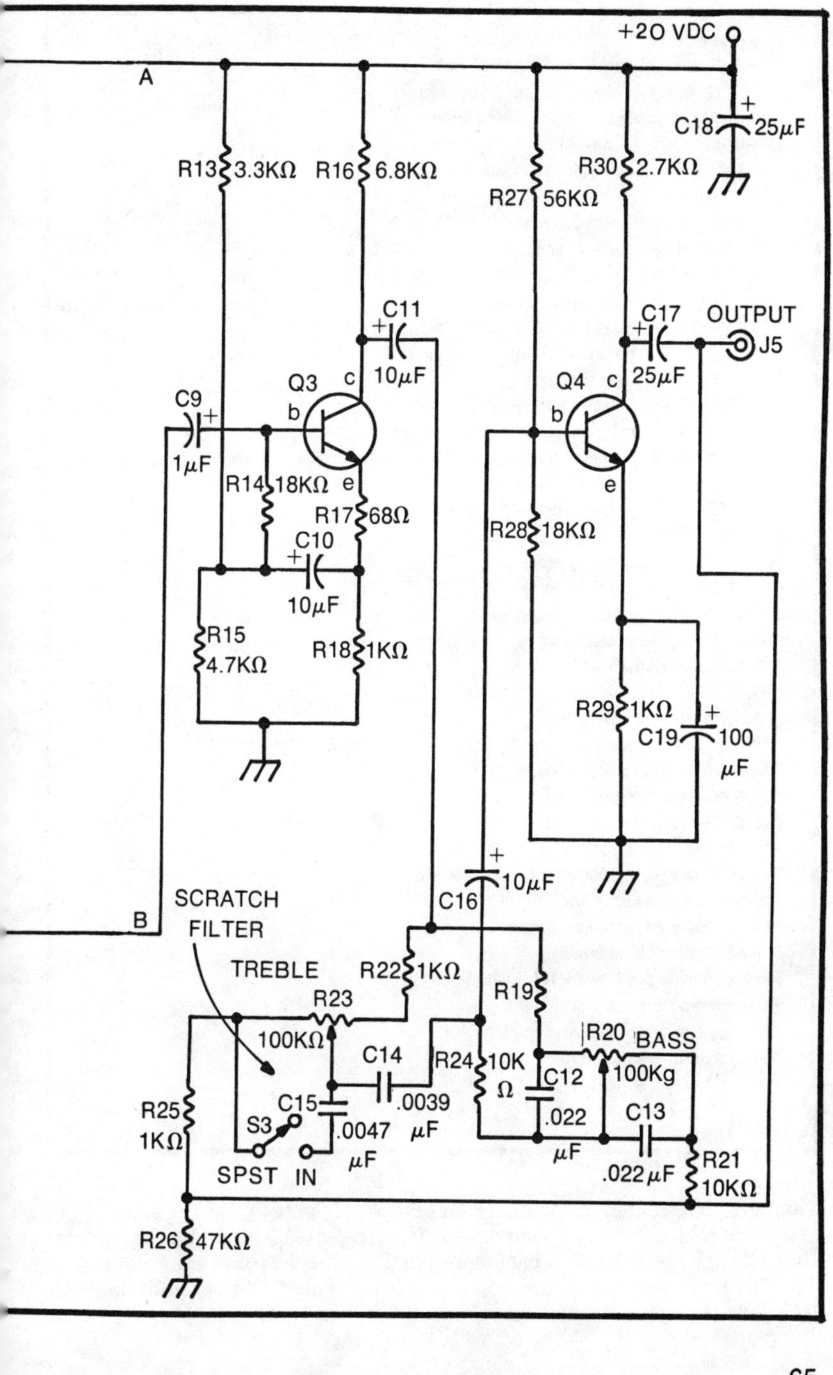
+20 VDC
A
B
C18 25μF
R13 3.3KΩ
R16 6.8KΩ
R27 56KΩ
R30 2.7KΩ
C11 10μF
C17 25μF
OUTPUT
J5
Q3
Q4
b
c
e
C9 1μF
R14 18KΩ
R17 68Ω
C10 10μF
R15 4.7KΩ
R18 1KΩ
R28 18KΩ
R29 1KΩ
C19 100 μF
C16 10μF
SCRATCH FILTER
TREBLE
R22 1KΩ
R23 100KΩ
R19
C14 .0039 μF
R24 10K Ω
R20 100Kg
BASS
C15 .0047 μF
S3
SPST
IN
C12 .022 μF
C13 .022μF
R25 1KΩ
R21 10KΩ
R26 47KΩ

PART LIST
C1—2 μF, 6-volt electrolytic capacitor
C2, C17, C18—25 μF, 25-volt electrolytic capacitor
C3—0.0027 μF paper or mylar tubular capcitor
C4—0.01 μF paper or mylar tubular capacitor
C5—100 μF, 3-volt electrolytic capacitor
C6, C11—10 μF, 25-volt electrolytic capacitor
C7—180 pF silvered mica capacitor
C8—0.033 μF mylar tubular capacitor
C9—1 μF, 12-volt electrolytic capacitor
C10, C16—10-volt electrolytic capacitor
C12, C13—0.022 μF paper or mylar tubular capacitor
C14—0.0039 μF paper or mylar tubular capacitor
C15—0.0047 μF paper or mylar tubular capacitor
C19—100 μF, 6-volt electrolytic capacitor
Q1—40233 general purpose amplifier transistor (Radio Shack 276-2009, or equal)
Q2, Q3, Q4—2N3242A general purpose amplifier transistor (Radio Shack 276-2009, or equal)
R1, R3, R16—6800 ohm carbon resistor
R2—180,000 ohm carbon resistor
R4—470 ohm carbon resistor
R5—27,000 ohm carbon resistor
R6—470,000 ohm carbon resistor
R7, R21, R24—10,000 ohm carbon resistor
R8—82 ohm carbon resistor
R9—1800 ohm carbon resistor
R10, R20, R23—100,000 ohm audio taper potentiometer
R11—8200 ohm carbon resistor
R12—250,000 ohm audio taper potentiometer (with tap) (Centralab F11-250K, or equal)
R13—33,000 ohm carbon resistor
R14, R28—18,000 ohm carbon resistor
R15, R31—4700 ohm carbon resistor
R17—68 ohm carbn resistor
R18, R22, R25, R29—1000 ohm carbon resistor
R26—47,000 ohm carbon resistor
R27—56,000 ohm carbon resistor
R30—2700 ohm carbon resistor
All carbon resistors 1/2 watt, 10% tolerance
S1—Single-pole, three-position rotary switch
S2—Single-pole, double-throw switch
S3—Single-pole, single-throw switch
Misc.—Chassis, perf board, pins

Fig. 6-10. Cont.

chapter. Looked at in this way, the unit is not so frightening as it seems at first. Note that R12 is different from most pots in that it has a tap. A manufacturer and part number are listed, but other manufacturers also make this item.

The wiring should be done on a perforated board and carefully laid out. The four transistor sockets are glued to the board and point-to-point wiring with terminal pins connects the components related to each transistor. The components should be wired directly to the pots in the third and fourth sections. This must be done before mounting the pots onto the panel of the metal chassis.

The power supply is 20 volts, single ended. It must be well filtered and regulated. Only 7.5 mA are required.

The inputs and the output leads should be separated from each other, but all should go on the rear panel. The front panel is used only for the controls and the ON-OFF switch, although the latter can be eliminated if the preamp is powered by the amplifier.

CHAPTER 7
COMPRESSOR AND CONTROLLERS

There is a world of listening difference between the professionally recorded tape and the one made at home. Without going into the technical differences, such as better microphones and an acoustically treated studio, the basic difference is in the better signal-to-noise ratio of the tape made by professionals. Not being able to record at a high level, the amateur keeps his recording gain down to avoid distortion due to sudden peaks in the music. Even the actor who suddenly raises his voice creates the same problem. Consequently, the recording gain is kept down in order to anticipate these peaks. But then, the resulting playback level is too low and consequently the tape and motor noises override the desired sound. The professional way is not to turn the gain down for the peaks, but to compress the highs so that an average recording level can be maintained. This method can be compared to lopping off high mountain peaks and raising the plateaus around them. You now have a high plain which is fairly uniform. Which is what the circuit to be described and illustrated in Fig. 7-1 attempts to do in a fairly simple way.

AUDIO COMPRESSOR

The unit is designed around the by-now familiar 741 op-amp. Note that the circuit is shown for one channel only. If you plan to record in stereo, the whole circuit should be doubled, but instead of the 741 IC use the 747, which is the dual version of the 741. Check the Appendix for the correct pin numbers for this IC.

Circuit

Whether you use the 741 or 747, no external compensation is required. These ICs have short circuit protection and consume little power. As to be expected, the input has high impedance with a low output impedance. The output voltage is quite constant, approximately 400 mV with an input signal ranging from 0.5 to 1.0 volt.

Sensitivity control R1 sets the point at which limiting begins. Voltage for the collector of Q2 is adjusted by means of R9, a trim pot, so that the voltage at

C1, C2, C5—1 μF, 25-volt tantalum electrolytic capacitor
C3—12 μF, 25-volt electrolytic capacitor
C4—4 μF, 25-volt electrolytic capacitor
D1—1N34A diode (Radio Shack 276-1102, or equal)
IC1—RS741 operational amplifier (Radio Shack 276-010, or equal)
J1, J2, J3, J4—Connector jack, to match existing equipment
Q1—2N5248 field effect transistor (Radio Shack 276-2036, or equal)
Q2—HEP629 general purpose amplifier transistor (Radio Shack 276-2007, or equal)
R1—100,000 ohm potentiometer
R2, R7, R8, R11—15,000 ohm carbon resistor
R3—56,000 ohm carbon resistor
R4—150,000 ohm carbon resistor
R5—1800 ohm carbon resistor
R6—68,000 ohm carbon resistor
R9—500 ohm, 1/4 watt trimmer potentiometer (screwdriver adjustable)
R10—5600 ohm carbon resistor
R12—2200 ohm carbon resistor
All carbon resistors 1/4 watt, 10% tolerance
S1—Single-pole, double-throw switch
S2—Single-pole, double-throw switch
Z1—6.8-volt, 500 mW zener diode (1N4736A, or equal)

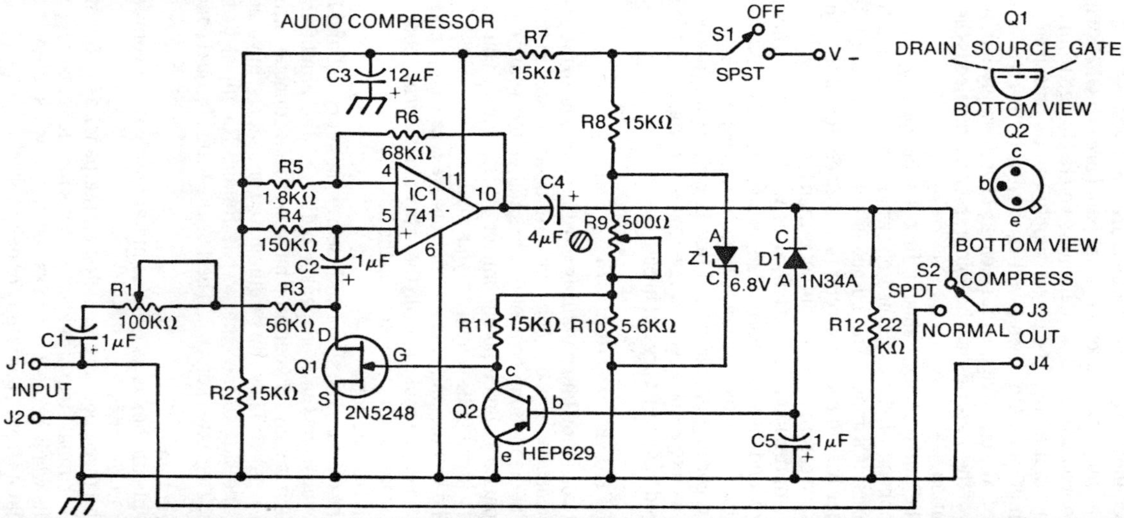

Fig. 7-1 Schematic for an IC and transistorized audio compressor. It is designed to reduce the high and increase the signal-to-noise ratio or a recording.

that point is 2.0 volts. It should remain within 10% of that figure. as that is the area at which Q1. a FET. works.

In the stereo version of the unit. it is most important to use a trim pot in each channel so that the channels are equalized.

The only controls on the front panel are ON-OFF switch S1. the sensitivity control and switch S2. which allows you to change from the compressing mode to a normal mode. The power supply can be batteries. which eliminates any possibility of hum. Two 9-volt batteries. connected in series. are all that's needed. If you have a well shielded 18-volt supply handy. you could use it instead of the batteries. as the compressor hardly needs the mobility.

Operation

The microphones should first be connected to a preamp (or to the Microphone Mixer described elsewhere in this book). then connected to the compressor's input. The output is connected to the tape recorder's inputs. Another way to use the circuit is to connect the tape recorder output to the input of the compressor. Its output is then fed to the amplifier's Tape In jacks.

Compression can be easily over worked. Experiment with the sensitivity control and use no more than is needed. Too much and the result is a monotone. Use just enough so that the highest peaks do not cause distortion. A good recording engineer. and that is what you are when you use this equipment. is like a good cook. A few spices are good. but too much ruins the dish. So. good taste must prevail in both cases.

HEADPHONE CONTROL BOX

If you enjoy late night listening to records or to the radio. but those around you do not share your hobby. you will be interested in the circuit shown in Fig. 7-2. The headphone control box has individual volume controls. one for each ear. plus a balance potentiometer which allows you to adjust comparative levels without getting up from your seat.

The volume and balance controls are wirewound pots. They should be low noise types. To make certain they stay quiet. spray them with a type of contact cleaner just before you close up the box.

Circuit Description

The circuit is so simple it is self-explanatory. R1 and R2 serve to attenuate the level from the loudspeaker outlets of the amplifier down to where the listening is easy. If your amp has a stereo headphone outlet. the resistors may not be needed. They can be eliminated or reduced in value. R3 shifts the sound from one side to the other. while R4 and R5 adjust to match varying input levels. These two pots would best be made up of a dual concentric unit.

Construction

A plastic box is as good as metal for a case. A convenient size is 3 3/4 × 2 5/8 × 1 1/2 inches.

The three pots are mounted on the panel as is the stereo output jack. A small terminal strip provides tie points for the remaining components.

The whole project is an easy evening's work. allowing you to eliminate all complaints from your family or neighbors.

REMOTE CONTROL FOR A HI-FI SYSTEM

You probably hate to get up from your chair to adjust the volume or balance of your amplifier. So. you either miss the weather report because the

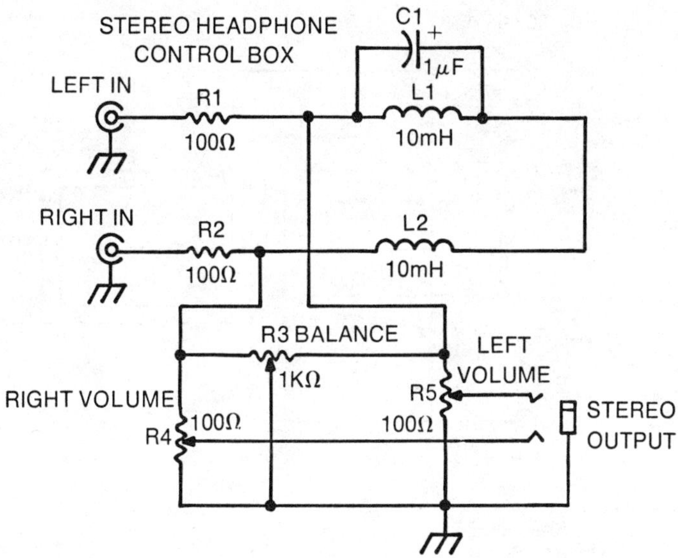

PARTS LIST
C1—1 μF, 10-volt electrolytic capacitor
L1, L2—10 mH inductor
R1, R2—100 ohm, 1/4 watt carbon resistor, 20% tolerance
R3—1000 ohm, 1/4 watt carbon resistor, 20% tolerance
R4, R5—100 ohm audio taper potentiometer
Misc.—Chassis (GC Electronics J4-725), 6-lug terminal strip, stereo headphone jacks

Fig. 7-2. Remote control "black box" for stereo headphones.

announcer is telling it confidentially or the crystal glasses are shattered by the screaming commercials. Without moving from your corner, you can now drop the volume or move the sound balance from one speaker to the other. The advantage of the circuit is illustrated in Fig. 7-3. You do not have to run shielded cable all over the room with the resulting high frequency loss. The control leads in this unit are isolated from the audio circuit.

The brain of this control is the Clairex Photomod. It combines a hermetically sealed photoconductive element with a neon lamp in a light tight encapsulated tube. By changing the voltage or current of the lamp the photoconductive element also changes. The higher the voltage or current, the lower becomes the output resistance.

Circuit Description

By increasing R3, the volume control, the light connected to each arm of the stereo channel is simultaneously receiving less voltage and as it dims the

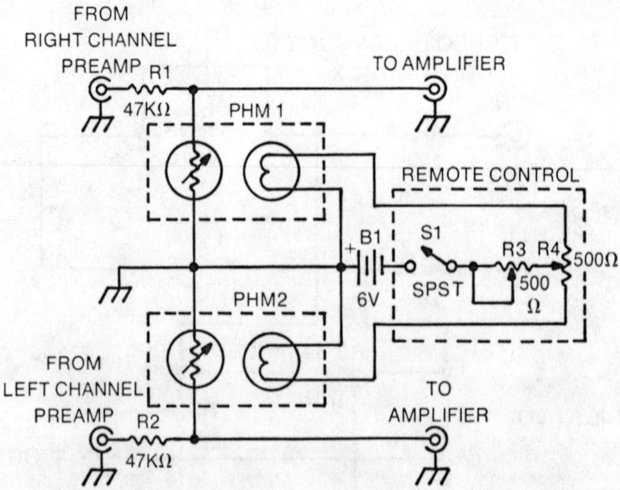

PARTS LIST
B1—6-volt lantern battery
PHM1, PHM2—Clairex Photomod CLM 5012
R1, R2—47,000 ohm, 1/2 watt carbon resistor
R3, R4—500 ohm, 4 watt audio taper potentiometer
S1—Single-pole, single-throw switch

Fig. 7-3. Schematic for remote control of amplifier. Circuit contains volume and balance controls.

higher becomes the resistance introduced into the signal path. Both channels will reflect exactly the same change. Balance control R4 reduces the resistance in one channel as it increases it in the other. The effect is always smooth and completely noiseless, since there are no mechanical switches or relays to add clicks and pops. A six-volt lantern battery will last some 180 hours.

Construction

A plastic or metal box needs to be large enough to hold R3 and R4, the two pots. The wires running to the Photomods could be zip cord as it is flexible and easily available. The total current carried in the circuit is only some 65 mA, so even a much thinner cord could do as well. The pots should be of a low noise type.

CHAPTER 8
MIXER CIRCUITS

The tape recorder "buff" knows the limitations of having only one or, at the most, two inputs as provided by most home recording units. To make professional sounding tapes, such as fading from one source to another, requires a mixer...because even two inputs are not enough to do a professional sounding recording. The mixer has many uses: feeding one tape recorder into another to which you are adding commentary to a piece of music, music as background to a play you might be taping, or simply background sounds to make a play come alive. A choral group or musical combo needs more than two microphones to get a quality recording. To record a singing guitarist requires one microphone for the instrument and one for the voice. To obtain the correct balance between the various voices and instruments, you must have control over many microphones and be able to feed each one with the correct level to the recorders.

PASSIVE MIXER

There are two types of mixers: active and passive. The passive circuit consists of a potentiometer and a resistor in each channel to separate one source from the other with no crosstalk. Figure 8-1 is a typical circuit for a passive mixer. As you can quickly see by examining the circuit, there is no signal gain. On the contrary, the needed separation results from using large value resistors with the consequent signal loss. There is only one advantage for that type of circuit: it is simple and cheap to build. There are many times when this type of circuit comes in handy, provided you can spare the drop in signal level. Also to recommend this passive circuit, or one similar, is that it requires a minimum number of parts for each input. If you have never used a mixer and want to find how it works, a passive unit will be fine for that purpose.

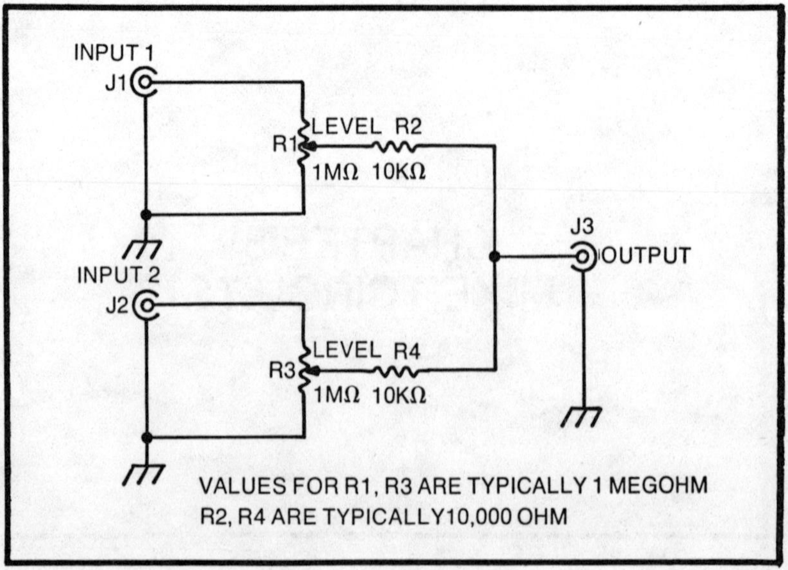

Fig. 8-1. Example of a passive mixer. Isolation is obtained at the cost of a lowered output level.

Let us say that you want to feed your record player into the mixer. If you have a ceramic cartridge, in spite of the signal loss, you may still get a signal through the mixer. But, if you own a magnetic cartridge, a passive mixer will be absolutely useless. A magnetic cartridge provides only 3.5 to 8 microvolts of signal. To hope to get any signal through the mixer, you would have to crank up the gain on the recorder all the way to make up for the circuit losses. This means a poor signal to noise ratio.

On the other hand, active mixers provide various levels of gain. In this case this may be only a unity gain—10 mV input providing 10 mV output. This is still better than a 0 mV output.

TRANSISTOR MIXER

The mixer shown in Fig. 8-2 is an active mixer with unity gain. Two microphones are connected to the mixer via J1 and J2. R1 and R3 set the input levels and, combined with R2 and R4, furnish the separation between channels. The combined signal is fed to the base of Q1, which amplifies the extremely low level signal resulting from the losses sustained due to the four resistors.

J1 and J2 are microphone jacks, either standard or the mini style, depending on the type of microphone you have. Q1 can be almost any general purpose silicon NPN transistor. The one chosen, HEPS0003, was picked merely because it was at hand. J3 could be a phono jack in which shielded cable leads to the input of the tape recorder. If you have a tape recorder with two inputs, two of the above mixers could be built, allowing you to use four microphones simultaneously. The microphone inputs could be on the front panel of a small metal chassis and the output on the back. The battery should go into the case with an ON-OFF switch on the panel. Keep all the leads as short as possible to avoid any danger of induced hum. Since R2 and R4 are mounted on their respective pots (as is C1), a tiny perf board will handle the remaining components.

Since the unit does not have any way of indicating whether distortion is occuring in the mixer due to levels being set too high, you must watch the VU meters on your tape recorder. If it does not have VU meters, it will have some other form of telling you when distortion begins, such as neon bulbs which flash on peaks. After some practice, the levels can be set in advance with confidence.

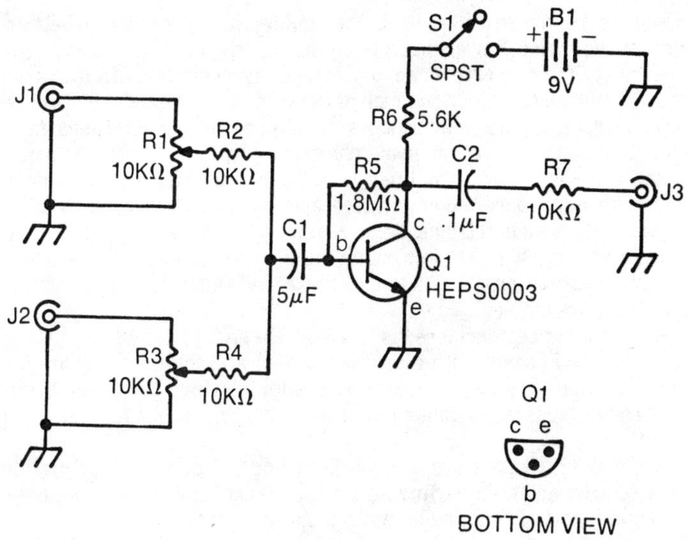

PARTS LIST

B1—9-volt transistor radio battery

C1—5 μF, 10-volt electrolytic capacitor

C2—0.1 μF mylar tubular capacitor

J1, J2—Microphone jack

J3—Phono jack

Q1—HEPS0003 general purpose silicon transistor (Radio Shack 276-2014, or equal)

R1, R3—10,000 ohm audio taper potentiometer

R2, R4—10,000 ohm, low noise metal film resistor

R5—1.8 megohm, 1/4 watt carbon resistor, 10% tolerance

R6—5600 ohm, 1/4 watt carbon resistor, 10% tolerance

R7—10,000 ohm, 1/4 watt carbon resistor, 10% tolerance

S1—Single-pole, single-throw switch

Fig. 8-2. A two-channel mixer using one transistor to achieve unity gain.

Only microphones were mentioned to be used with the mixer. Ceramic phono cartridges could also be mixed for the reason mentioned above. The best microphone to use with this mixer is a high impedance dynamic type.

AN IC MIXER

The professional mixers are expensive and with reason. The quantity of transistors with their biasing resistors and bypass capacitors add up to a lot of parts—and dollars, not to mention the labor in assembling such units. They have many frills that the amateur can well do without.

The mixer described here is an active mixer with adequate gain for magnetic cartridges as well as for dynamic microphones. It makes use of an IC.

ICs offer many advantages to the builder and designer, whether professional or hobbyist. Fewer discrete components are used, thereby reducing the possibility of wiring errors; not to overlook the economy. Circuit variations are greatly eliminated, since so much of the circuit is within the body of the IC. Until fairly recently, only transistors were used to replace tubes in mixer circuits. Transistors have low noise characteristics, an important asset in a mixer, as noise is equally amplified with the signal. ICs, on the other hand, when first introduced were comparatively noisy, a factor negating their other advantages in an audio circuit. Several linear ICs have now made their appearance offering low noise, low harmonic distortion...and some are even internally short-circuit proof. This makes these well suited for our purpose. And, their price is reasonable.

One such IC to be considered is the RCA CA3048. It is a 16-pin dual-in-line silicon integrated circuit with four independent and identical AC amplifiers. All four operate from a single-ended power supply, which can range from 12 to 16 volts. The latter is the absolute maximum. See Fig. 8-3 for the illustration of the circuit.

Each of the high gain amps has a high impedance non-inverting input—typically 90 kHz, and an inverting lower impedance for the feedback network. Internal and external coupling between the amplifiers is reduced by the use of two positive and two ground terminals. Noise is low, only 2 dB at 1 kHz. The output AC voltage is typically 2 volts rms.

The gain of the amplifiers is largely determined by the amount of feedback resistance. This gain may range from a high of 60 dB, with only a wire for the feedback, to a low of 15 dB with a 100K ohm resistance feedback. For the experimenter who wishes to vary the gain to suit himself, the table in Fig. 8-4 will suggest varying resistances, varied gains and maximum inputs. There are two ways to obtain the precise gain for each input. The first is to use at least 5% tolerance resistors for gain consistency. The other is to use an audio generator and a millivoltmeter. For the actual resistor use a minipot with a value of about 10K. Make sure the pot does not introduce noise into the circuit.

Construction

This is definitely a PC board project and is not meant to be a first time project, as it is complicated. However, the beginner could tackle it if he wishes to approach it using a "bite-size" technique. Regard each amplifier on the CA3048 as a separate circuit. That makes it easier to plan the layout of the PC

(POWER SUPPLY SHOULD BE WELL FILTERED AND REGULATED)

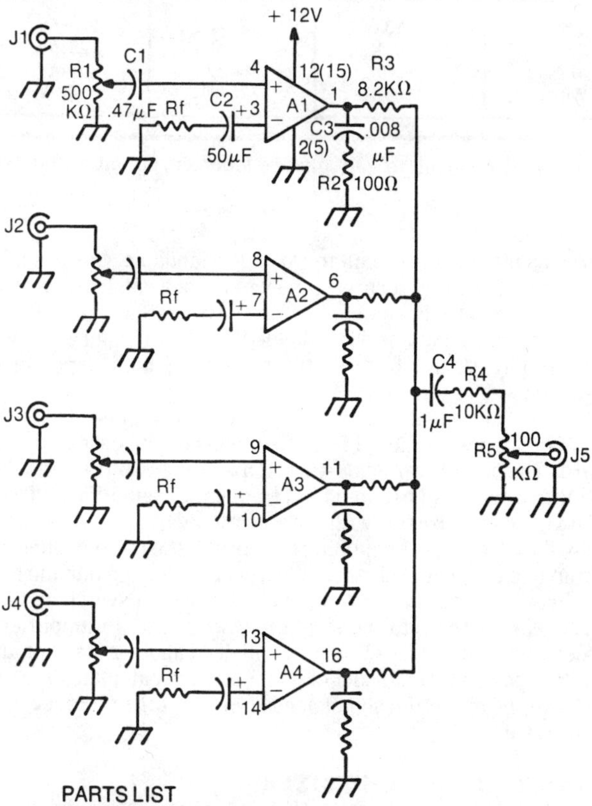

PARTS LIST

A1, A2, A3, A4—CA3048 (RCA) quadruple amplifier IC

C1—0.47 μF mylar tubular capacitor

C2—50 μF, 3-volt electrolytic capacitor

C3—0.008 μF mylar tubular capacitor

C4—1 μF, 25-volt electrolytic capacitor

J1, J2, J3, J4—Microphone jack

J5—Phono jack

R1—500,000 ohm, audio taper potentiometer

R2—100 ohm, 1/4 watt carbon resistor

R3—8200 ohm, 1/4 watt carbon resistor

R4—10,000 ohm, 1/4 watt carbon resistor

R5—100.000 ohm, audio taper potentiometer

Fig. 8-3. An IC with a four-channel input is used to serve as a mixer in this circuit.

Rf in ohms	Gain	Upper Limit	Maximum Input
5.6K	10X	35 kHz	225 mV
330	100X	70 kHz	25 mV
47	400X	80 kHz	5 mV
Wire	1000X	100 kHz	3 mV

Fig. 8-4. Table shows values to vary gain of feedback circuit for unit shown in Fig. 8-3.

board. as well as making the circuit more understandable. One great help is to breadboard the CA3048. It is surprising how much one can learn about making signal paths and a general layout.

The power supply should be well shielded. due to the high gains within the circuit. The supply should be well filtered—using high capacitance electrolytics—and be well regulated.

Inputs and outputs should be kept away from each other. as in any circuit of that type. Shielded wire is used from the inputs on the case to the inputs on the PC board. The output terminals—if you are using two ICs—should be presented with the proper load to avoid high frequency oscillations. (Two ICs are mentioned if you plan to use this with a stereo recorder.)

As you will notice. when examining the circuit. the microphones cannot be switched from one channel to another. Whatever is plugged into the right-hand board will always be heard in that channel. But. with a switch on the output. both channels can be fed into the single mono input of an amplifier. Unless your recorder is extremely sensitive. you will need another stage of amplification. This will depend on the amount of gain you have calculated for your feedback circuit and the minimum signal needed in your tape recorder to give you a good clean signal.

FOUR CHANNEL MICROPHONE MIXER

The unit shown in the photograph (Fig. 8-5) is a highly sophisticated mixer. It's more complicated to build. but it offers much more flexibility than the other mixers described before and it allows a quantity of options to be added.

The prototype as photographed did not have the VU meters on the front panel. Instead. they were outboarded so they could be closer to the director during the recording sessions. On the photo. rotating pots are shown. Later. the front panel was changed to accommodate slide pots. They give an instant visual indication of the level at which that pot has been set. In the lower left corner. a switch had originally been put in to permit the headphones to switch from one channel to another. The switch was omitted later and stereo headphones were incorporated. In spite of these changes. which are only cosmetic. the guts and the operation of the mixer remained the same.

Figure 8-6 shows the layout suggested. based on a chassis of 1579 inches. But. this will be discussed later in the construction section of the project.

Description

From the block diagram (Fig. 8-7). you will note that four inputs and two outputs are shown. More inputs can be added. if desired. The two outputs can

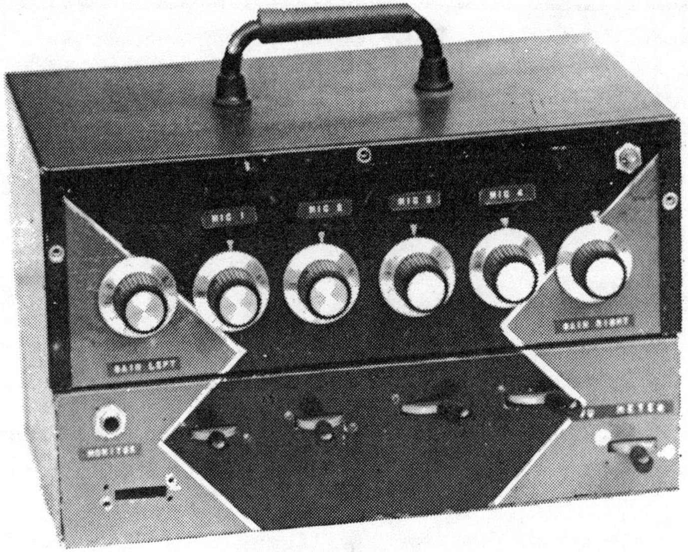

Fig. 8-5. Prototype of mixer.

be tied together to enable them to feed one channel of an amplifier, which would then have all four inputs. A third output is available at J7.

Each signal amplifier is designed to accept two inputs. They are all identical. The two master amplifiers, which are contained on one PC board, are slightly different from the signal amplifiers. But, the same PC board pattern can be used. The wiring is slightly different, that is all. This is a time where it would be worthwhile to make the PC boards with the photographic process. The difference between the master and the signal amplifiers is obvious when

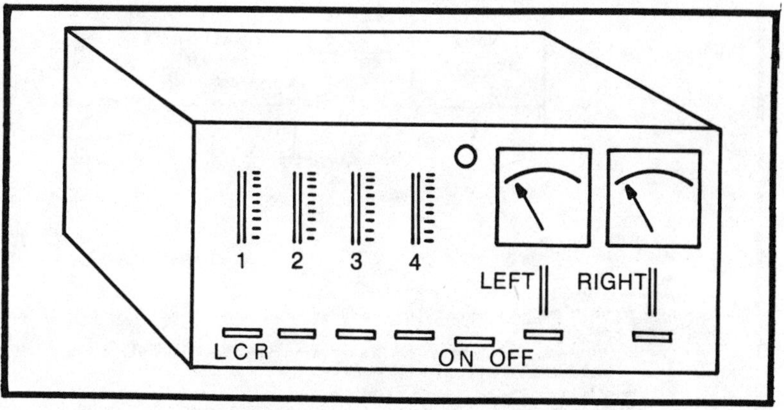

Fig. 8-6. Sketch for idealized front panel of a 4-channel mixer.

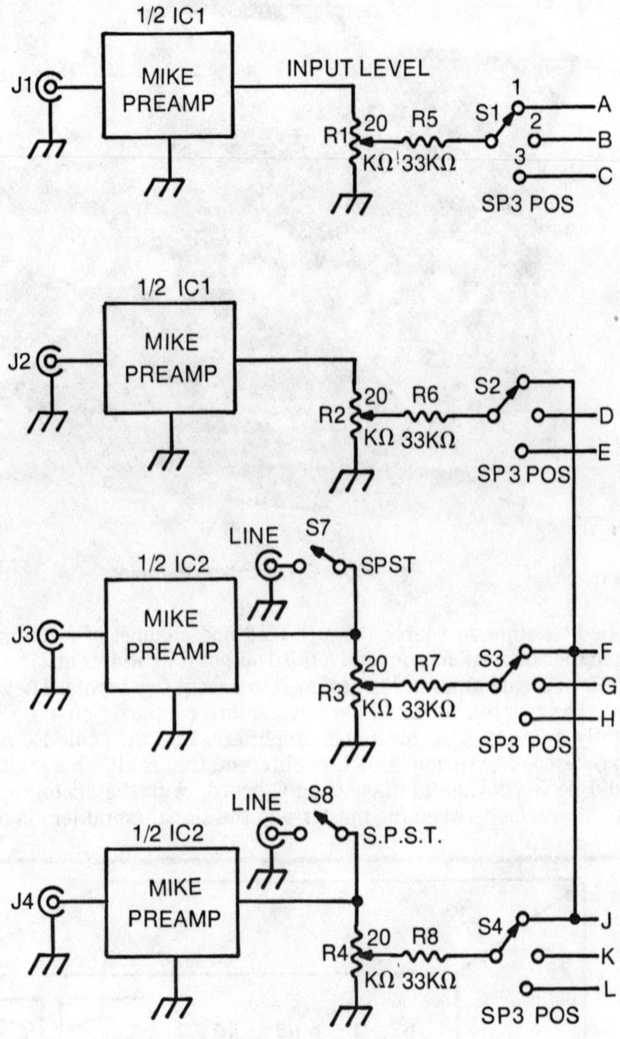

PARTS LIST
J1, J2, J3, J4—Microphone jack
J5, J6, J7—Phono jack
J8, J9—Headphone jack
R1, R2, R3, R4—20,000 ohm, audio taper, slide actuated potentiometer (GC part number B1-702, or equal)
R5, R6, R7, R8—33,000 ohm, 1/2 watt carbon resistor, 10% tolerance
R9, R10, R11, R12—10,000 ohm, audio taper, slide actuated potentiometer (GC part number B1-702, or equal)
R13, R14—3300 ohm, 1/2 watt carbon resistor, 10% tolerance
R15, R16—68,000 ohm, 1/2 watt carbon resistor, 10% tolerance
S1, S2, S3, S4, S5, S6—Single-pole, three-position switch (Switchcraft S3036L, or equal)
S7, S8—Single-pole, single-throw switch

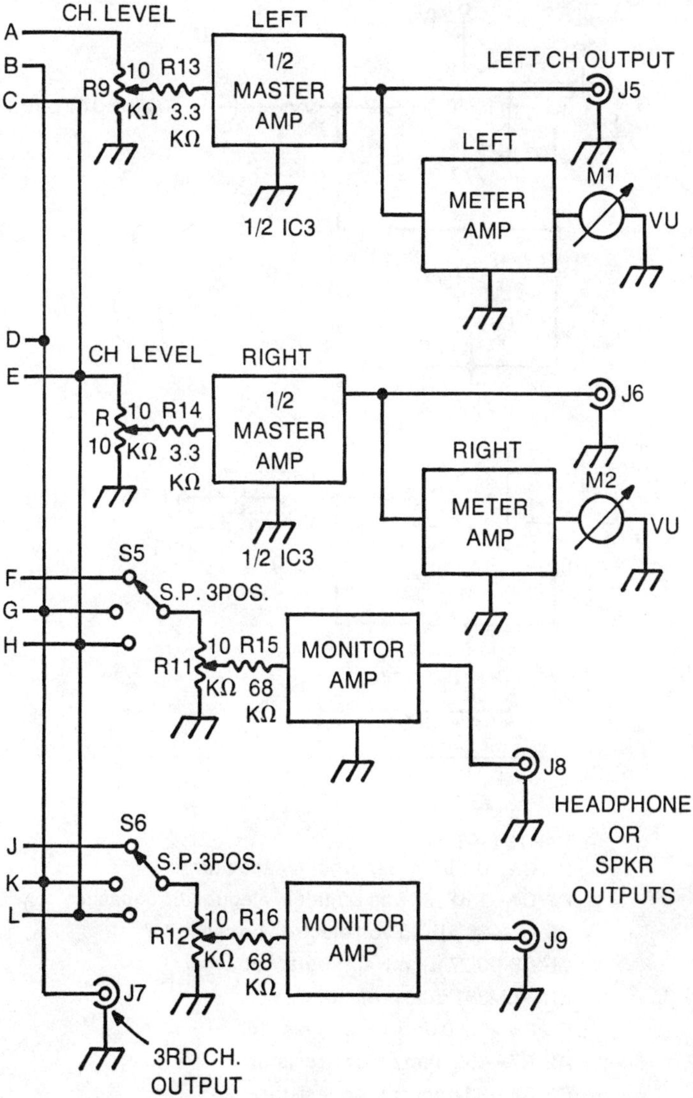

Fig. 8-7. Block diagram for a 4-channel mixer. Note extreme flexibility obtained by the switching arrangement.

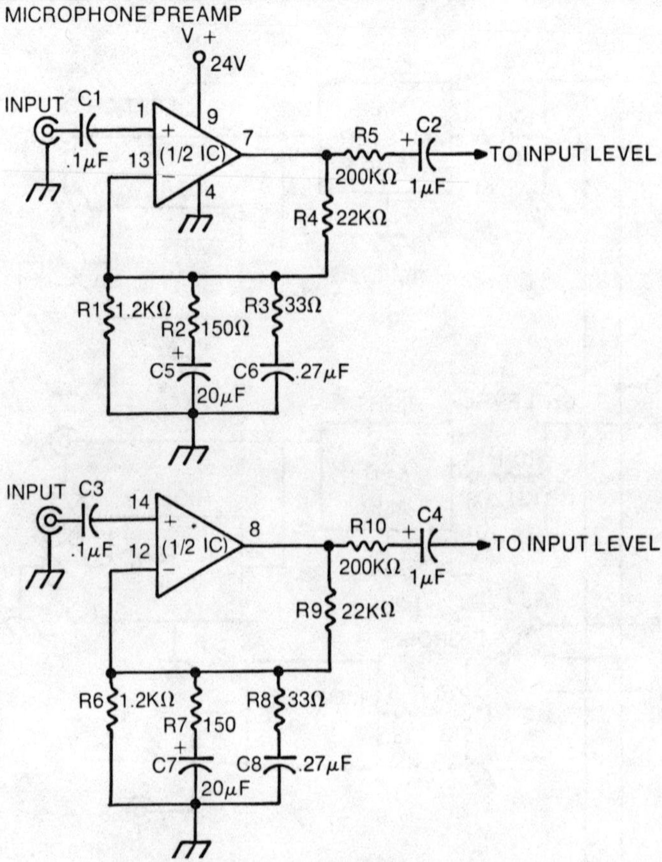

PARTS LIST

C1, C3—0.1 μF mylar tubular capacitor
C2, C4—1 μF, 35-volt tantalum electrolytic capacitor
C5, C7—20 μF, 35-volt electrolytic capacitor
C6, C8—0.27 μF mylar tubular capacitor
IC1—LM381 dual preamp IC
R1, R6—1200 ohm carbon resistor
R2, R7—150 ohm carbon resistor
R3, R8—33 ohm carbon resistor
R4, R9—22,000 ohm carbon resistor
R5, R10—200,000 ohm carbon resistor
All resistors 1/4 watt, 10% tolerance

Fig. 8-8. Schematic for microphone signal board.

you compare Fig. 8-8 and Fig. 8-9. The equalization is different with five less components for the master amplifier board.

The signal enters at J1 through J4. and is amplified by half of the stereo IC. R1 to R4 are the input level controls. The levels are further reduced by R5 through R8. This attenuated level enters switches S1 to S4. In position 1 of each switch. the signal is diverted to the right channel. In position 2. the signal goes to the cue bus. In position 3. the signal is fed to the left channel.

The signal then appears at R9 or R10. the master channel levels. This output is dropped by the pots and resistors R13 and R14. The remaining signal. which appears at the output jacks J5 and J6. is metered by the right or left meter amplifier with its respective meter. (This is an optional feature as you can use the meters on your tape recorder.)

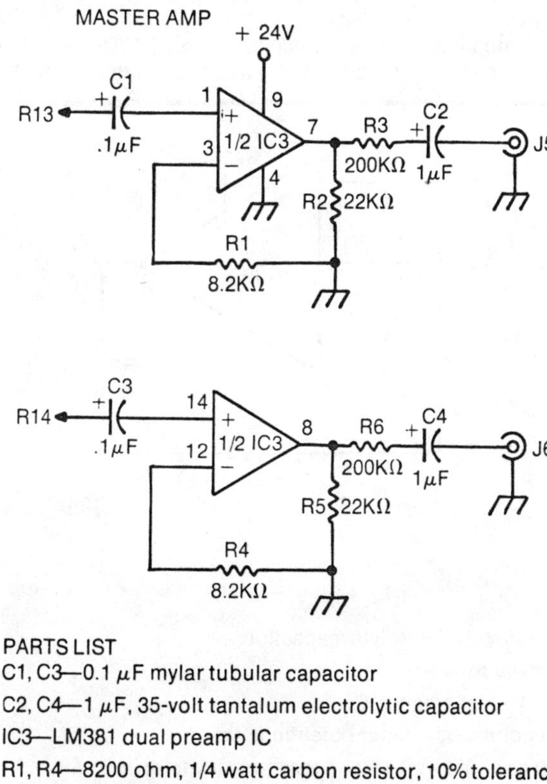

PARTS LIST
C1, C3—0.1 μF mylar tubular capacitor
C2, C4—1 μF, 35-volt tantalum electrolytic capacitor
IC3—LM381 dual preamp IC
R1, R4—8200 ohm, 1/4 watt carbon resistor, 10% tolerance
R2, R5—22,000 ohm, 1/4 watt carbon resistor, 10% tolerance
R3, R6—200,000 ohm, 1/4 watt carbon resistor, 10% tolerance

Fig. 8-9. Schematic for master amplifier boards. Both amplifiers are on the same board, but are entirely separate.

Each channel has its own monitor amplifier with its own jack (Fig. 8-10). This, however, could be a stereo jack for stereo headphones or two small speakers. The speakers could only be used if the recording is taking place in another room or else you will get an echo effect, which you may not want. R11 and R12 adjust the listening levels.

The third channel, the cue bus, permits adding a third signal source by means of J7. Since this channel does not have its own amplifier one would have to be added. The cue bus offers another advantage in that you can check the level of a signal source, by means of S5 and S6, before switching it into either of the other two channels. This can even be done during a recording, since the suggested switches are noiseless. This also means that a voice can be moved from the right to the left, or vice versa, with no clicking noise appearing in the recording.

Signal Amplifiers

Each signal amplifier consists of one-half of an IC. The one chosen is the LM381, a dual-in-line stereo preamp. It is easily obtainable by the hobbyist at

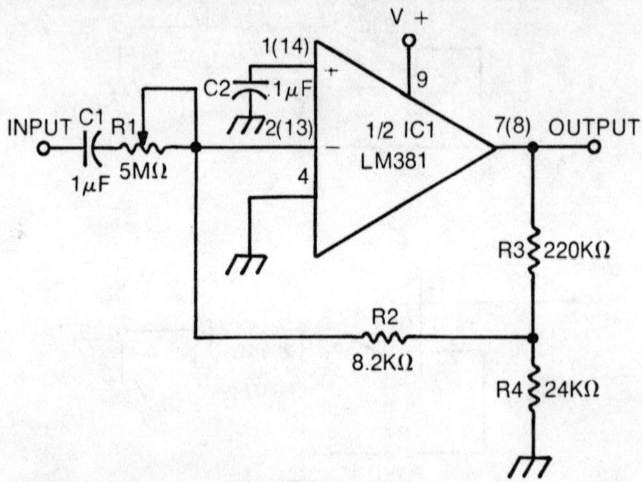

PARTS LIST

C1—1 μF, 35-volt electrolytic capacitor

C2—0.1 mylar tubular capacitor

IC1—LM381 dual preamp IC

R1—5 megohm audio taper potentiometer

R2—8200 ohm, 1/4 watt carbon resistor, 10% tolerance

R3—220,000 ohm, 1/4 watt carbon resistor, 10% tolerance

R4—24,000 ohm, 1/4 watt carbon resistor, 10% tolerance

Fig. 8-10. Circuit for the single channel of a monitor amplifier, which is an optional feature of the mixer. Notice the resemblance to the microphone signal board. Same PC pattern can be used, if desired.

most mail order houses and it is carried by GC Electronics, who sell through Heath, as well as Radio Shack and Lafayette Radio. The LM381 is made expressly for low noise applications. It has an internal power supply decoupler-regulator. There is 60 dB of channel separation at 1 kHz, eliminating the possibility of crosstalk. Another of its features is its high gain (typically 112 dB). However, in the circuit shown, the gain has been reduced to 60 dB. The LM381 is not fussy about voltage; it will operate with anything from 9 to 40 volts from a single-ended supply. For its use in the mixer circuit, 24 volts was chosen.

To show the simplicity of the signal amplifier, it was breadboarded, as you will see in the photograph, Fig. 8-11. The equalization network consists of R1, 2, 3, 4 and C3 and C4. In the schematic for the master amplifier (Fig. 8-9), this network is replaced by R2. With this one difference, the signal amplifiers become master amplifiers. In the microphone preamp, R5 and C2, with the equivalent components in the other channel, are not wired on the PC board. This is also true for R3 and C2 on the master preamp.

A circuit for a magnetic phono preamp is shown in Fig. 6-2. Again, the same board could be used with slight modification. It is the equalization that is different. The full description of the magnetic preamp is in the Chapter 6 preamps.

If you plan to mix a line signal, such as a radio or another tape recorder, you would bypass the signal preamp. The reason for this is that the signal from a radio requires no preamplification before being mixed. Switches S7 and S8 can be added on the back panel.

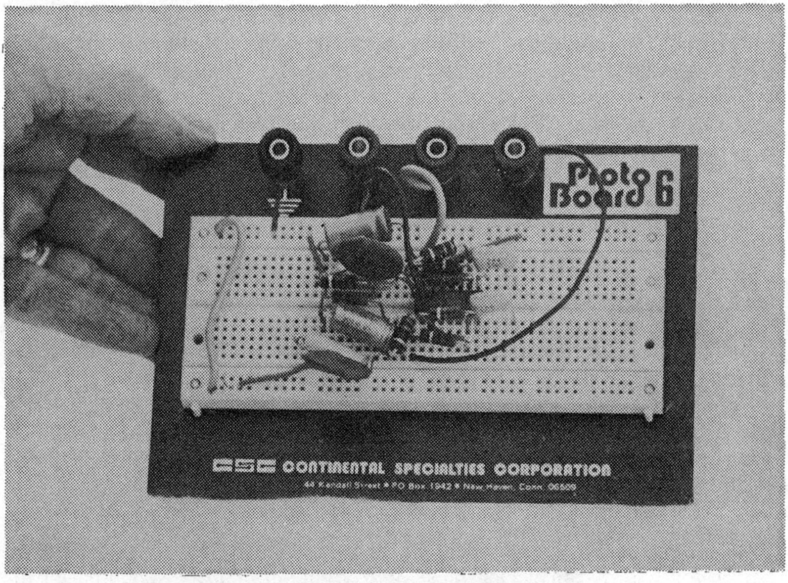

Fig. 8-11. Breadboarded microphone preamp. Breadboarding is used to optimize a circuit and it is also of help in making the PC pattern.

The mixer was designed for use with high impedance microphones. These are the type usually found with home recording units. Should you want to use low impedance microphones, a line transformer plugged into the jack at the mixer makes the needed transformation. You can also use the Impedance Matcher described in another chapter.

VU METER

Some means of observing recording levels to get high signal to noise ratio and obtain distortion-free results is needed. Even with a tape recorder equipped with AGC (Automatic Gain Control), the upper frequency limits may be reduced, but when the input signal is too low to begin with, nothing can be done except to start recording over again. "Magic eyes" were used on home tape recorders back in tube days (some even had two), which at least told you if you were within the right ball park when you were recording. Now most units use VU meters. However, with an add-on mixer, it helps to have a meter in the mixer circuit itself, or even two to indicate what is going on in each channel. It also is helpful in getting well balanced recordings.

Light Emitting Diodes (LED's) are being used in some units to indicate if the signal is too low, within the right levels, and if too high...high enough to cause distortion. But, let's consider the way the professionals do it in a recording studio—with VU meters. The mix-down consoles, used by rock recording groups, have some 16 inputs and each has its own meter. For our purpose we'll be satisfied to put one meter for each channel—right and left—rather than one for each input. That way each meter measures the total gain for that channel.

If you have a recorder in which the recording bias can be adjusted, the VU meter comes in handy to regulate the bias to suit various tape formulas. The procedure is to record a 1 kHz tone at a level which is indicated by 0 dB on the meter when the recorder is operating at 7.5 ips. Without changing any controls on the recorder, measure the output as you play back. If the bias is set correctly for that tape, the result should also be a 0 dB reading. If not, the bias should be adjusted to obtain that result. The same method will also allow you to choose the best tape for your machine.

Description

Each VU amp (Fig. 8-12) consists of a grounded emitter stage which provides minimum loading of the circuit to which it is connected. For the transistor, almost any NPN silicon audio amplifier transistor will do. Even if the two transistors do not match, it does not matter. The volume control for each meter can be adjusted so the readings approximate each other. A steady signal of one volt rms output from the mixer should indicate a value of 0 dB. Once this setting has been arrived at, the pots can then be forgotten.

In the absence of a high impedance AC voltmeter, you can record a steady tone and adjust the meters on the mixer so they agree with those on your tape recorder. This does not mean that the recorder's meters are necessarily accurate, but it is better than nothing.

Construction

The easiest way is to mount the few components on a small piece of perforated board with wiring pins. Both amplifiers can be mounted on the same board. At least, this is true if both VU meters are going to sit next to each other in the front panel. The board is then fastened to the meters' terminals. (See Fig. 8-13.) It is better in actual practice to have both meters next to each

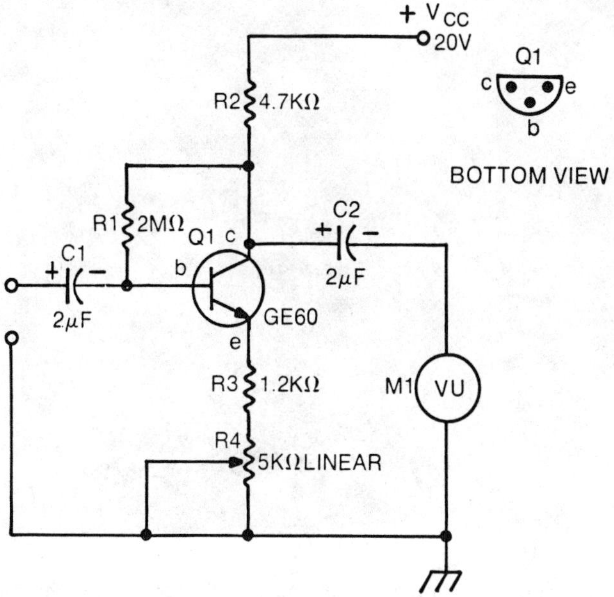

PARTS LIST
C1—2 μF, 5-volt electrolytic capacitor
C2—2 μF, 35-volt electrolytic capacitor
M1—VU meter (CG part number D1-930, or equal)
Q1—GE60 silicon general purpose audio transistor (Radio Shack 276-2013, or equal)
R1—2 megohm, 1/2 watt carbon resistor, 10% tolerance
R2—4700 ohm, 1/2 watt carbon resistor, 10% tolerance
R3—1200 ohm, 1/2 watt carbon resistor, 10% tolerance
R4—5000 ohm linear taper trimmer potentiometer

Fig. 8-12. VU meter circuit for mixer. Two of these are needed.

other, as this avoids glancing from one on the right to the one on the left as in a tennis match. The two pots should be of the mini-type that can be fitted into the perforations in the board.

VU meters (a completed unit is shown in Fig. 8-14) indicate only the general circuit response, so extreme accuracy is not as important as though volts or amperes were being measured. The pots should be mounted in such a way so that they can be adjusted with a screwdriver while a tone is being recorded.

In the event of a lack of a signal generator and voltmeter, you can set the meters by recording a few sentences spoken in an even tone. On the playback

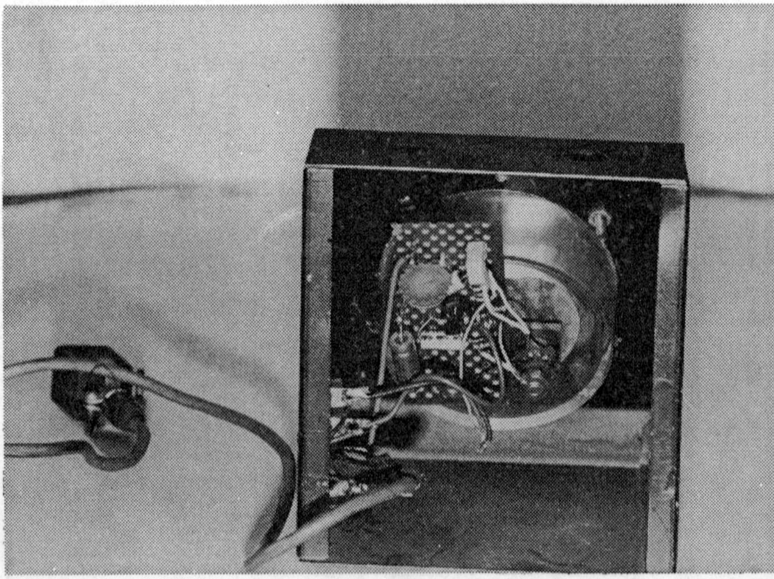

Fig. 8-13. Showing mounting of circuit board on the terminals of the meter. This unit was designed for outboarding at a distance from the mixer.

Fig. 8-14. The VU meter with its own amplifier. This unit can be kept at a distance of some two feet away from the mixer. This can be an advantage in some cases.

Fig. 8-15. Inside view of mixer. The power supply components are mounted at the extreme right of the photo.

of the tape, you can hear if distortion has resulted and set your meters accordingly. This is an extremely rough and very far from satisfactory way, but it is a solution of a kind. An internal view of a completed mixer is shown in Fig. 8-15.

CHAPTER 9
TIMER CIRCUITS
BASED ON THE 555 IC

The 555 integrated circuit is not exclusively an audio device. It can be employed to generate tones in the audio range, as you will see demonstrated in the toy organ project that follows. Moreover, it can be used as a switch. This IC is a flip-flop or multivibrator. It contains a highly stable controller able to produce extremely accurate time delays or oscillations. The 555 can be used for precision timing, with the timing intervals ranging from as little as 10 microseconds up to 10 seconds. Even this maximum interval can be stretched to a timing period lasting several days by hooking up several of these ICs in sequence. Originally manufactured by Signetics, many other manufacturers are producing them and every mail order house has them listed in their catalog.

But, that's not all this circuit can do. With only a few external components, it can be made to generate square waves with varying duty cycles over a wide frequency range, convert sine waves to square waves without changing their frequency...and as frosting on the cake, generate triangle waves. It may even be used to make a simple pulse generator.

With all of these qualities, the unit is made even more useful for the hobbyist by being very tolerant of supply voltage. It can operate on as little as 5 volts and as high as 18 volts. Although it is best to use it in the 5- 15-volt range, for the reasons mentioned before when we discussed the upper limits of other ICs. In spite of its lower 5-volt operating extremity, nevertheless the author used the 555 as a switch and, wanting to make it very lightweight, found it could work on three AA cells.

As can be seen from Fig. 9-1 the internal circuitry of the 555 is complex, but in actual practice there are only eight pins to which external components are connected. This is a great circuit to breadboard, as you may find novel uses for this IC that have as yet not been discovered. Study Fig. 9-1 to develop new projects.

The reason for the detailed explanation of the 555 is that only by understanding its "innards" can you put this device through its paces intelligently. In the

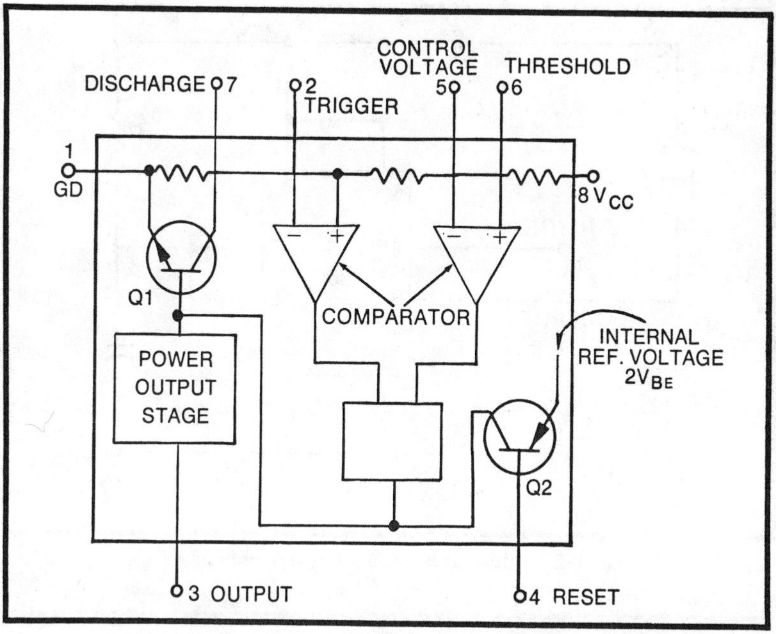

Fig. 9-1. Block diagram of internal circuit of the 555 IC.

monostable mode, the 555 acts as an accurate timer...the timing period being set by a combination of only two components, R1 and C1 in Fig. 9-2. The external capacitor connected between pin 7 and ground is initially held discharged by the short circuit path of Q1 inside the timer. Applying a negative pulse to pin 2, the flip-flop changes state, which releases the short circuit and drives the output high. C1 now charges through R1 from the positive supply. The time constant associated with R1-C1 determines the rate at which C1 charges. When the voltage across the capacitor equals two-thirds of the supply voltage, the threshold comparator resets the flip-flop and discharges C1, driving the output to the low state. It also provides an output pulse at pin 3.

Once triggered, the circuit will remain in that state until the preset time has elapsed, even if another triggering pulse were to be applied during this interval. Since the charge rate and the threshold level of the comparator are both directly proportional to the supply voltage, the timing interval remains independent of the supply.

By means of this extremely simple arrangement, timing periods ranging from 10 μsecs to 10 seconds may be obtained. See Fig. 9-3 for the values of C1 and R1 to arrive at any needed timing interval within the above range.

We mentioned the repeatability and stability of the 555. However, as you can understand, the timing cycle is affected by the leakage of the capacitor and the tolerance of the resistor in the timing circuit. With those limitations, the IC is an excellent base on which to design a variety of timing devices.

SIGNALING CIRCUIT

The circuit illustrated in Fig. 9-4 shows the use of the 555 as a signaling device. In this case a Sonalert, which is a solid state buzzer, will sound once trig-

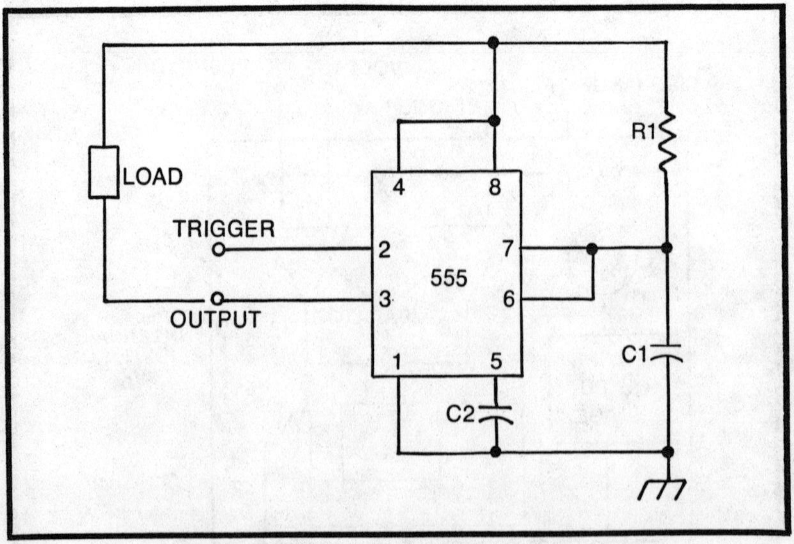

Fig. 9-2. Circuit for the 555 in the monostable mode.

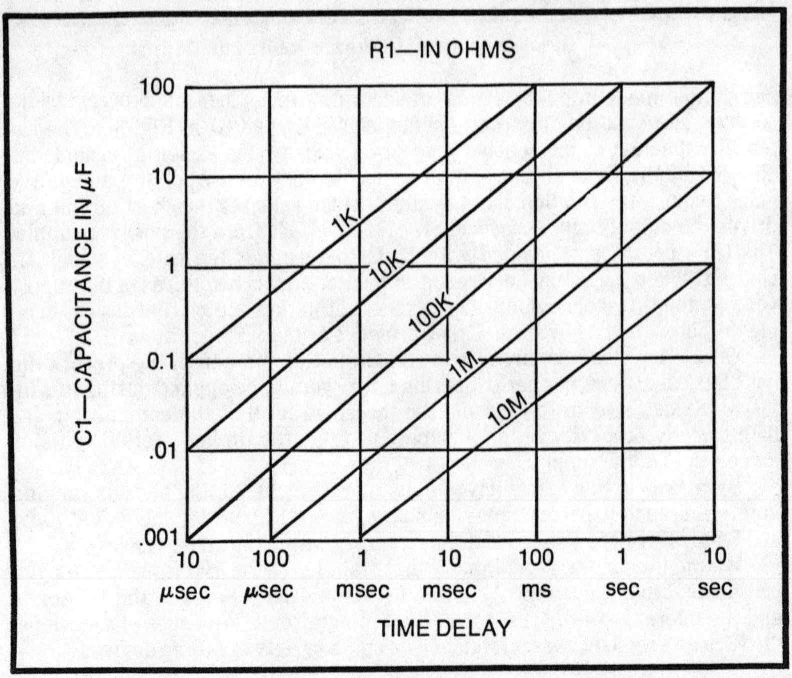

Fig. 9-3. Graph to help calculate time delays from 10 μsec to 10 seconds.

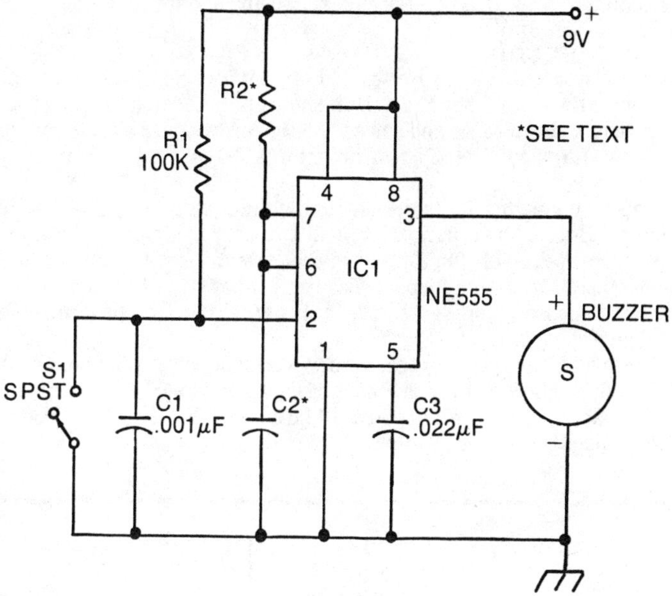

PARTS LIST
C1—0.001 μF disc ceramic capacitor
C2—Calculated for desired delay, see text
C3—0.022 μF disc ceramic capacitor
IC1—NE555 timer IC (Radio Shack 276-1723, or equal)
R1—100,000 ohm, 1/4 watt carbon resistor
R2—Calculated for desired delay, see text
S—Sonalert SC628 solid-state buzzer
S1—Single-pole, single-throw switch, of mechanical type choosen to suit application need

Fig. 9-4. 555 employed as the control in a signaling device. The IC drives a solid state buzzer which eliminates relays and mechanical sounding devices.

gered, and will continue sounding as long as the time delay calculated from Fig. 9-3 exists. Switch S1 applies a negative pulse to pin 2. The switch, which should be a momentary close type, could be a manual switch or it could be a touch plate under a rug. C1 and C3 are used to keep the circuit from oscillating. The Sonalert is made in a variety of models. The one chosen for our use is the SC628. This model produces a 2900 Hz tone and the unit operates on 6 to 28 volts DC. It consumes only 14 mA maximum.

RELAY SWITCHING

Instead of the Sonalert a relay could be used instead to control a higher voltage than can be handled by the 555 directly. Figure 9-5 shows such an arrangement. The relay does not have to be particularly sensitive, since the current output of the IC is generous. The relay should have a rating of about 12 volts and drawing some 100 mA. The reason for diode D1 in parallel with the relay coil is to prevent a reverse emf from being applied to the 555.

VISUAL INDICATOR

If a visual indicator were wanted, an LED could be used instead of the Sonalert or the relay. (See Fig. 9-6). It should be noted that an LED does not require more than 1.6 volts and since the 555 operates over a wide supply range, a resistor R3 should be used to limit the LED current and drop the extra voltage.

This by no means limits the uses of the timer chip. It can be used as a rain indicator. Here's how: on a board, nail two copper strips so that they do not quite touch. A drop of rain falling between the strips will close the circuit and sound a buzzer or a light indoors. The two copper strips are connected, as you may have guessed, to pins 1 and 2. In this case, the drop of rain replaces the switch in Fig. 9-4.

From all of this you can see just how versatile the 555 can be, and its many applications. There is only one note of caution: the 555 can withstand many supply voltage variances, but you must be careful that the resistor between pin 7 and the supply is never less than 1000 ohms.

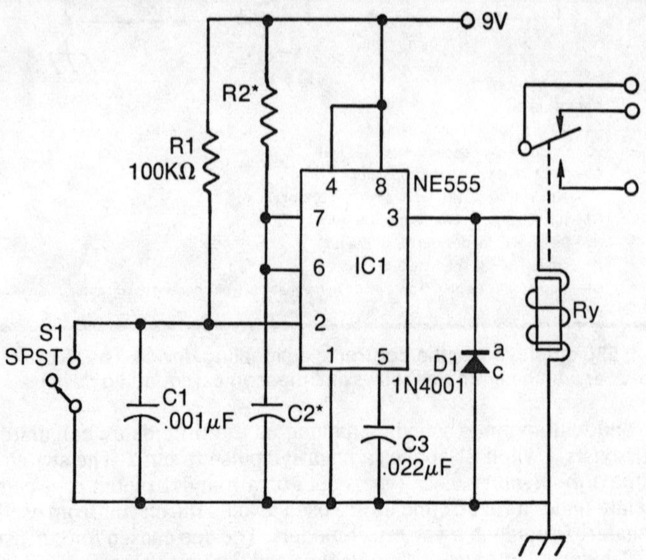

PARTS LIST
D1—1N4001
Ry—See text
Other component values are similar to Figure 9-4

Fig. 9-5. Circuit in which the IC controls a relay to switch loads, which could not be handled by the 555 directly.

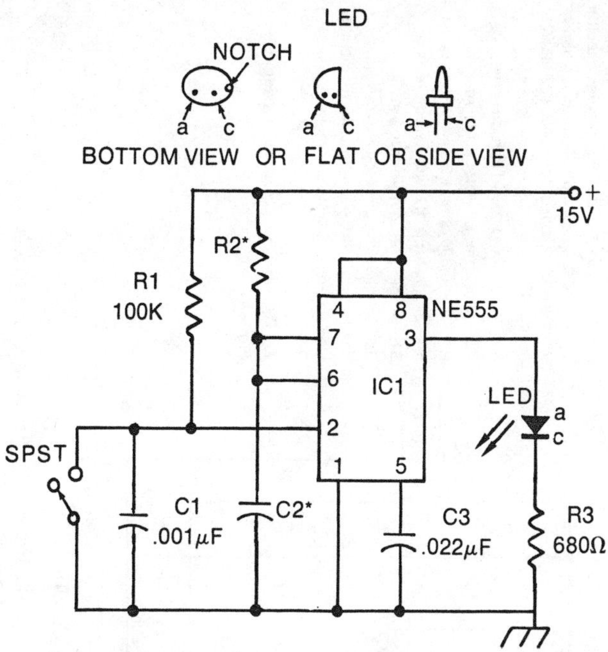

PARTS LIST
LED—Light Emitting Diode (Radio Shack 276-041, or equal)
R3—680 ohm, 1/4 watt carbon resistor
Other component values similar to those in Figure 9-4

Fig. 9-6. An LED is used as a visual indicator in this circuit, which uses the 555 as the timer.

TOY ORGAN

The instrument described below and illustrated in Fig. 9-7 is meant for a child to pick out melodies and is not a replacement for the mighty theatre organs of days gone by. The unit is monophonic; only one note can be played at a time with a range of 10 notes worked out. You can add more if you wish. All you have to do is make up additional combinations of Rx and Ry with a switch for each.

The circuit resembles what you have already seen for the 555. In the test section of the book, this IC is made to operate as a square wave generator. For the organ, it is also used to generate square waves with their rich harmonic content. But, with an additional transistor and a total of three parts, triangle waves can also be generated. These are even richer in harmonics.

The frequency of a tone is derived from the formula:

$$\text{Frequency} = \frac{1.44}{[2.2K + 2(Rx + Ry)]C1}$$

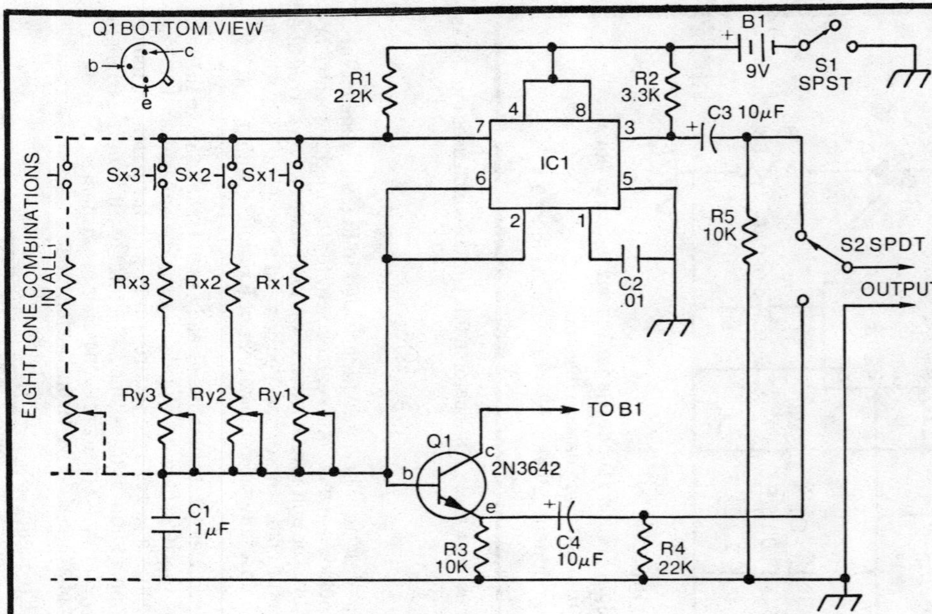

PARTS LIST

B1—9-volt transistor radio battery

C1—0.1 μF mylar tubular capacitor

C2—0.01 μF mylar tubular capactor

C3, C4—10 μF, 15-volt electrolytic capacitor

Q1—2N3642 general purpose silicon audio amplifier transistor (Radio Shack 276-2009, or equal)

R1—2200 ohm, 1/2 watt carbon resistor

R2—3300 ohm, 1/2 watt carbon resistor

R3, R5—10,000 ohm, 1/2 watt carbon resistor

R4—22,000 ohm, 1/2 watt carbon resistor

S1—Single-pole, single-throw switch

S2—Single-pole, double-throw switch

SX1 through SX8—Single-pole, momentary ON push-button switch, or metal key (see Fig. 9-9)

For Rx and Ry values, see Frequency Chart, Fig. 9-8

Fig. 9-7. A toy organ with a 10-note range, illustrating the flexibility of the 555. It is used here to produce tones with pre-set frequencies.

The equation can be reduced (to be solved more easily) by converting to:

$$Rx + Ry = \frac{7.200.000}{Fx} - 2200$$

When the IC is used as a square wave tester, it is operated as a free running multivibrator. This means that once the power is turned on, the tone (square wave) is continuous until the power is interrupted. For our present purpose, it is hooked up so the tone continues only as long as the switch, or key, is depressed and held.

Circuit Description

The IC is operated as an astable multibrator with approximately a 50% duty cycle, the frequency being determined by the tone combination of Rx, Ry, C1, and R1. Each tone uses a momentary ON switch, a fixed resistor (R1), Rx, and a trimmer pot (Ry).

At the output, pin 3 of the 555, a 9-V P-V square wave is available. The coupling capacitor (C3) is provided to remove the 4.5 volt bias at this point.

Q1, and its associated components, R3, R4, and C4, furnishes a different set of harmonics for each tone, resulting in a different sound. This feature is optional. Q1, as an emitter-follower, is used to buffer the voltage across C1 and provide a low impedance from this high impedance point.

Switching S2 from one position to another furnishes a note, which though it is of the same frequency, sounds very different due to the added harmonics.

The frequencies remain independent of the power supply voltage. The tone will be reduced in volume as the battery wears down, but that will be the only effect of lowered voltage.

Only approximate frequencies have been calculated in Table 9-1. Due to the 10 to 20% tolerances of the resistors and the capacitor, you will be only within a fairly close area to the tone wanted. This is the reason for Ry, the trim pot, in each tone combination. The easiest way to calculate other frequencies is to choose Ry so that it is half of Rx. Then, take the nearest value obtainable in a trim pot.

Construction

Once the components are assembled—remember you need Ry, Rx, and S for each tone. A piece of perf board with the printed circuit type of trim pot al-

Table 9-1. Chart to Calculate Approximate Frequencies For Each Tone Of The Organ.

Frequency	Note	Rx + Ry (K)	Rx (K)	Ry (trimmer)
261.63	C	25.4	18	10
293.66	D	22	18	10
329.63	E	20.3	18	10
349.63	F	18.4	10	10
392.00	G	16.2	10	10
440.00	A	14.2	10	5
493.88	B	12.4	10	5
523.25	C′	11.57	8.2	5
587.33	D′	10.06	8.2	5
659.26	E′	8.7	5.6	5

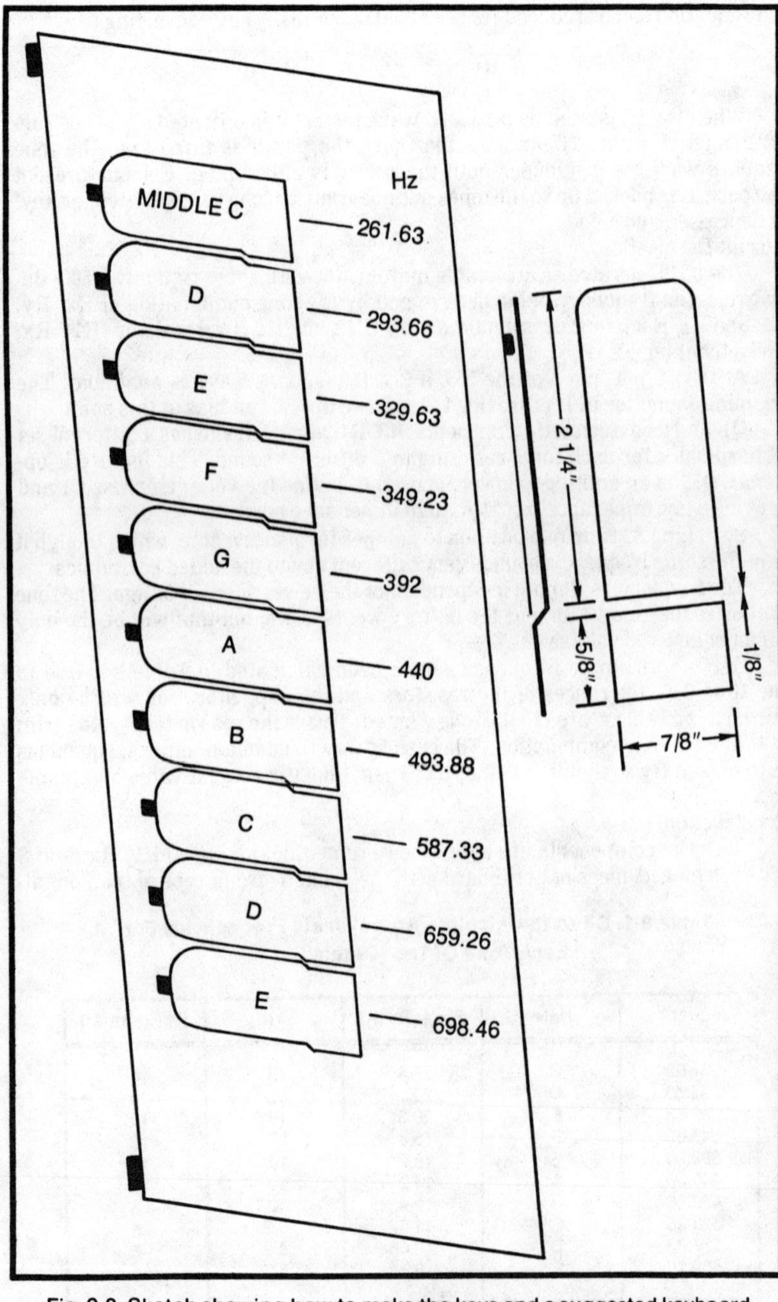

Fig. 9-8. Sketch showing how to make the keys and a suggested keyboard.

lows you to mount them on the board itself. They must be accessible for alignment. After that they can be forgotten. The whole thing can be mounted inside a metal chassis whose size depends on the number of tones you want. From each tone combination, extend leads to the switches or keys.

To make the organ look realistic, a key board can be made from scrap tin as in Fig. 9-8. File the edges smooth. As the key is depressed, the lead from that tone combination makes contact with the common lead to pin 7. Just recently, organ keys were offered in some mail order catalogs which would simplify the construction and improve the looks of the unit.

Alignment

First, take the note "A" as an example. Set the trim pot at about 4K. Combined with R2 (10K), you are fairly close to the 14.2K needed. C1 never varies. Get hold of a piano, or other instrument that is in tune, and slowly adjust the trimpot for that tone until the note beats with the standard. Do the same with the other notes. Once you are satisfied with the results, put a drop of varnish on the pots so they do not change in adjusted value.

The output level of a 555 not being sufficient to drive a loudspeaker, a small amplifier can be connected to the output. As a means of keeping peace with the rest of the household, headphones can be used instead of the loudspeakers...at least until a certain playing expertise has been arrived at.

CHAPTER 10
MISCELLANEOUS CIRCUITS

This chapter presents some "fun" circuits, which are easy to build, and which will make your electronic entertainment more enjoyable. Such circuits are a lie detector, a surf sound producer, a color organ, a white noise generator, a Theremin, and an experiment for communicating with light. Also included are more serious devices, such as an electronic metronome and a phase lock loop stereo demodulator circuit.

PARTY LIE DETECTOR

Have you ever wondered whether the person who answers a question of yours is telling the truth? You probably have at some time or another. So do the police when they are questioning a suspect. That is why the polygraph, or lie detector, as it is popularly known, was invented. The unit described here is a lie detector—of sorts. But, whereas the polygraph is fairly dependable, if it is handled by a skilled operator, and even admitted in some courts, our little gadget is strictly a party game and not to be taken seriously. On a continously moving chart, the polygraph indicates heartbeat, evenness of breathing and response, increased sweating, and several other factors which are all taken in consideration when analyzing the honesty of the one being questioned. It is the combination of all these factors which make the polygraph a serious testing instrument. Its principle is based on the idea that when a normal person—not a psychopath who cannot tell the difference between a lie and the truth—tells a lie many internal things happen to that person. Their heart will beat faster, their response to a question in which they will lie will be somewhat slower, but above all they will begin to perspire more. Our unit measures only the decrease in skin resistance due to sweating caused by an attempt to lie. So, our unit is meant only as a game; it will start off a party with a real bang. However, to avoid breaking up the party before it begins, it would be wise to avoid questions thay may cause a person embarrassment.

The basis for the party lie detector is simple. When we become emotional due to pain or danger, the glands go to work and release perspiration all over the body. Maybe the reason for that was that in our primitive days it helped to slither out of an enemy's grasp if we were sweat covered.

When this sweat appears on the person connected to the probes of the lie detector, the skin resistance drops and the needle on the meter of our unit shoots up. We are not concerned how much it shoots up, just the fact that the person is not reacting normally to our question.

A Wheatstone bridge is the heart of the lie detector. See Fig. 10-1. Wheatstone bridges are used to measure not only resistances, but capacitances as well...and very accurately. In the example shown, we have four resistances forming the arms of the bridge. R1 and R2 form one arm while the combination of R3 and R4 make up the other arm. If voltage is applied to points C and D, current will flow through the two arms; the amount of current being dependent on the values of the resistances. If all the resistors are equal, the current through one arm will obviously be equal to that in the other. But even if they are not alike, as long as the sum of the resistances in one arm is equal to the sum in the other arm, the bridge is not unbalanced and will show no current flow. For example: if R1 is 100 ohms, R2 is 120 ohms, while R3 is 80 and R4 is 140, the value of each arm is a total of 220 ohms.

To prove this, we go back to Ohm's Law. Let us assume that 100 volts are applied to the bridge. Since R1 equals 100 and R2 equals 120, then the current flowing through the resistors is equal to the voltage applied to those resistors divided by their resistance.

$$I = E/R$$
$$= 100/220 = 0.45 \text{ amperes}$$

Since the arm made up of R3 and R4 has the same total resistance, the current flow is identical. The bridge being balanced, the meter registers nothing.

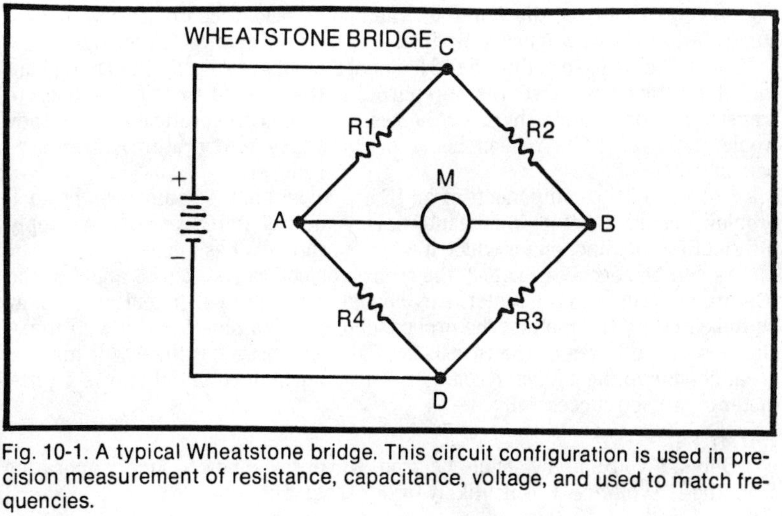

Fig. 10-1. A typical Wheatstone bridge. This circuit configuration is used in precision measurement of resistance, capacitance, voltage, and used to match frequencies.

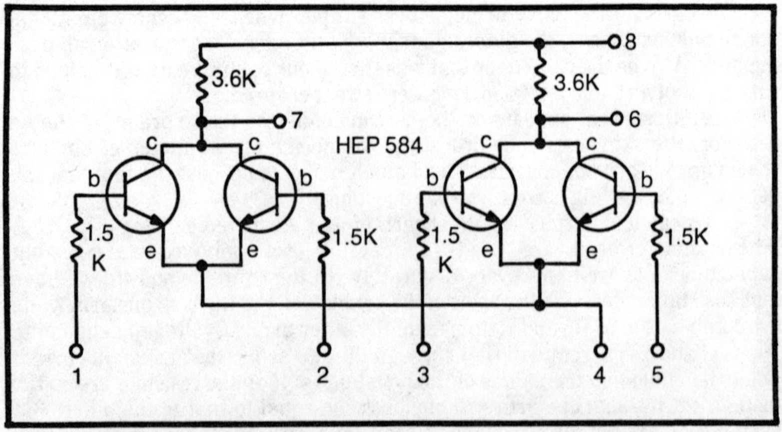

Fig. 10-2. Block diagram of the HEP584, a dual 2-input gate.

Should there be a change in the situation, i.e., should any one of the arms have a different resistance, the voltages at points A and B will no longer be equal. The meter will then indicate this change.

Let us assume that branch AD is now made up of a known resistance—the skin resistance through a pair of probes balances the bridge. If a pot is placed on that arm it can be adjusted so that R1 and R2 balance the bridge. This is true only when the person being questioned is at ease. Once the victim becomes nervous, the skin resistance drops, the bridge becomes unbalanced and the meter indicating a current flow. The amount of current is unimportant. It is not the quantity, but the drop of skin resistance that is of interest to the interrogator.

When you compare the circuitry within the IC in Fig. 10-2, schematic in Fig. 10-3, you will see the familiar Wheatstone bridge configuration, but instead of resistors we are using two transistors for the arms of this bridge.

The IC being used is the HEP584, a dual 2-input gate digital IC. Though not meant for this use, it serves our purpose. If we were to employ discrete transistors, substituting them for the resistors in our theoretical bridge, they would not work. The reason being drift, due to temperature differences between them.

Since all of the components of an IC are all within the same case, no such problem occurs. By studying the internal circuitry of an IC, we can often apply the unit to some function for which it was not originally designed.

As can be seen in Fig. 10-3, the transistor collectors—pins 6 and 7 of the IC—are connected to the meter. Any change between their potential will now be indicated by the meter. The probes, connected to pins 1 and 5, will make changes in the bases of the transistors. The tiny base variance will make a great change in the collector voltages. This change in potential allows a 1 mA meter to be used successfully.

CAUTION

Only a 9-volt battery should be used. Never use a power supply operating from the 117-volt line. It is not likely to short, but still.....

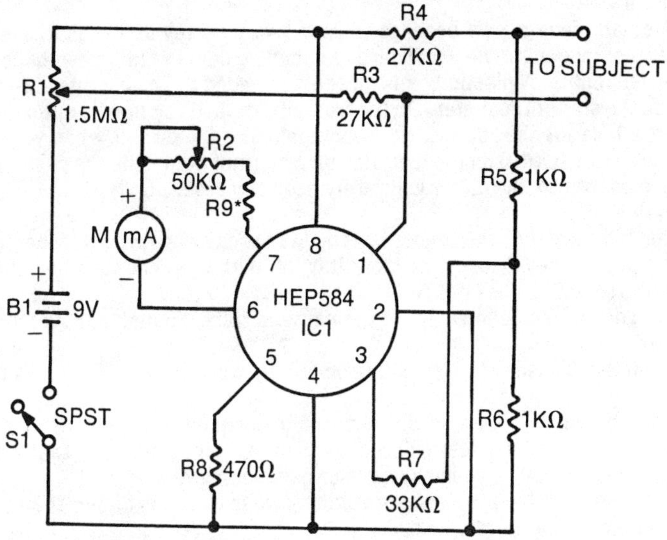

*SEE TEXT

PARTS LIST
B1—9-volt transistor radio battery
IC1—HEP584 dual 2-input NOR gate integrated circuit
M1—0-1-0 milliampere meter (100 ohm internal resistance)
R1—1.5 megohm linear taper potentiometer
R2—50,000 ohm linear taper potentiometer
R3, R4—27,000 ohm, 1/2 watt carbon resistor, 20% tolerance
R5, R6—1000 ohm, 1/2 watt carbon resistor, 20% tolerance
R7—33,000 ohm, 1/2 watt carbon resistor, 20% tolerance
R8—470 ohm, 1/2 watt carbon resistor, 20% tolerance
R9—See text

Fig. 10-3. A lie detector using the 584. Circuit employes a bridge configuration by making use of the transistors inside the IC.

Construction

Probes should be made from two small pieces of tin or soft aluminum which fit around the wrists and held tightly in place with straps or rubber bands. They are connected to the case by means of flexible wire, the type used for test leads.

The meter circuit made up of R2 and R3 is flexible enough to permit the use of most current meters. R9 is important, because if R2 is reduced to zero you would have enough current flowing through the meter to burn it out. The value of R9 may change to suit your particular meter. A center-zero meter is better as it allows the needle to swing in either direction. But, a 0-1 mA meter will work well.

A small piece of perf board is as easy a way as any on which to assemble the few components. The perf board can then be held in place by means of the meter terminals. A plastic box is all that is needed for a case, its size dependent on the size of the meter. On the front of the case you need a cutout for the meter, a hole for the shaft of R1, which balances the circuit when you begin, and the switch to turn on the unit. R2 can be a mini-pot inside the case. Once it is adjusted to zero, the meter will rarely need any resetting.

Operation

The "victim" has the probes fastened to his or her wrists. Now begin your questioning as they do in crime labs. Start off with innocent questions that can be answered with "Yes" or "No".

"Is your name John Doe?" If the name is right, no meter movement will be seen.

"Is today Tuesday?" Depending on the answer, there might be no needle movement.

Then comes the trap in the form of a pertinent question. But, even a truthful person under stress will react—even to their name—and that must be taken into account when conducting your questioning.

Don't become too personal, especially if your victim is bigger than you, or has a larger husband or boyfriend!

ELECTRONIC METRONOME

This portable metronome is useful for any musician to maintain his beat, but is equally useful in the photographic darkroom as it produces loud clicks. The unit can be calibrated so that setting a dial tells you how many clicks per second will be heard. There will be no variances later, so that the calibration can be permanent. The timing is independent of battery voltage. The level of the clicks will drop off as the battery is exhausted, but that is the only effect of lowered battery voltage. See Fig. 10-4.

Q1 is a unijunction transistor operating as a relaxation oscillator. Q2 raises the output level. With no voltage, Q1 is at ground potential since the capacitor is discharged initially. When voltage is applied, current flows through R4 and the voltage at the transistor rises. But, the voltage quickly drops, due to C1 being discharged.

When Q1 is not conducting, the base of Q2 is relatively negative so no current flows. When Q1 is pulsing, Q2 is switched on and the pulse is now heard through the loudspeaker, which acts as the collector load for Q2.

The time constant of the circuit depends on the setting of R2 and of R1-C1. With the component values in the circuit as shown in the parts list, clicks will be heard from four per second to one per four seconds. For less frequent beats, increase the value of C1.

Wiring should be no problem. A plastic case with holes burned in for a 2-inch speaker and for the shaft on R2 are all that is needed. For long life, obtain a mercury battery to avoid changing too often.

SOUND OF THE SURF

The sound of the surf coming from your amplifier when you connect the circuit shown in Fig. 10-5 will be enough to get you to put on your bathing suit.

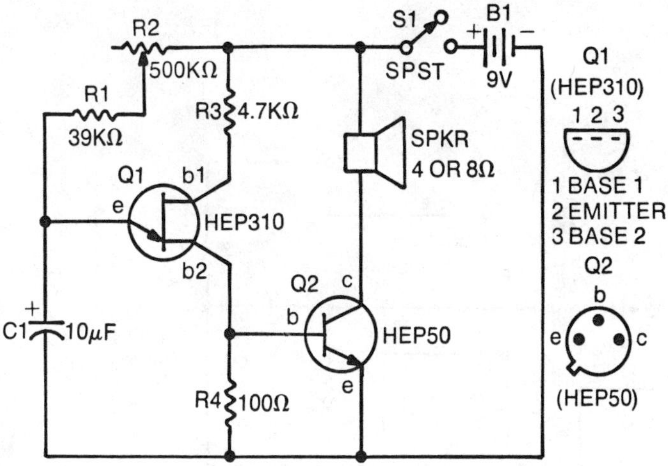

PARTS LIST

B1—9-volt transistor radio battery

C1—10 μF, 12-volt electrolytic capacitor

Q1—HEP310 general purpose unijunction transistor (Radio Shack 276-2029, or equal)

Q2—HEP50 general purpose audio amplifier transistor (Radio Shack 276-2009, or equal)

R1—39,000 ohm, 1/2 watt carbon resistor, 20% tolerance

R2—500,000 ohm, audio taper potentiometer

R3—4700 ohm, 1/2 watt carbon resistor, 20% tolerance

R4—100 ohm, 1/2 watt carbon resistor, 20% tolerance

SPKR—4 or 8 ohm speaker

S1—Single-pole, single-throw switch (may be mounted on R2)

Fig. 10-4. A portable metronome for a musician or for timing in the darkroom. The entire unit is self-contained.

Add Hawaiian music and you are all ready for a genuine luau. The basis of this realistic surf sound is the creation of "white noise". This is produced by the reverse biasing of Q2. Notice that the collector is not connected. This produces a sound which is random in frequency and amplitude.

The white noise is fed to Q3 whose base is coupled through R8 to the collector of the phase shift oscillator Q1. Q3 conducts and ceases at the same rate as does the potential at the collector of Q1. This causes the noise from Q3 to rise and fall at the same rate. The phase shift oscillator takes a total of some 10 seconds to run the complete cycle.

Potentiometer R10 can be put in the circuit temporarily to make certain that the sound is not cut off entirely, but instead drops to a low level, before rising again. Once you have determined the right amount of resistance to get the effect you want, you can either put a dab of varnish on the pot so the adjusted value does not change or substitute a fixed resistor of the same value.

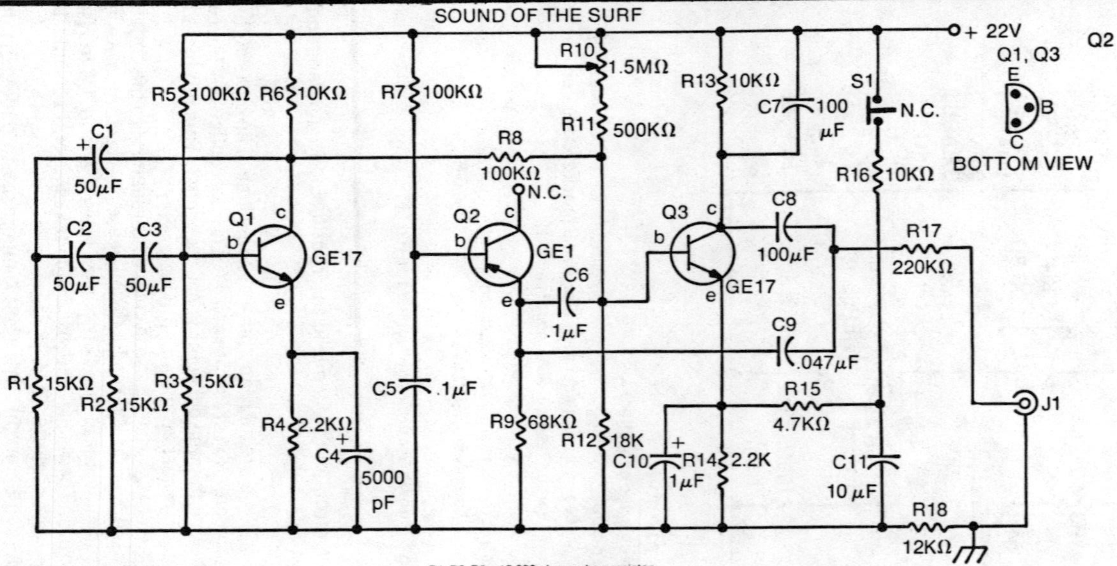

PARTS LIST

C1, C2, C3—50 μF, 35-volt electrolytic capacitor
C4—5000 pF disc ceramic capacitor
C5, C6—0.1 μF disc ceramic capacitor
C7, C8—100 μF, 35-volt electrolytic capacitor
C9—0.047 μF disc ceramic
C10—1 μF, 12-volt electrolytic capacitor
C11—10 μF, 12-volt electrolytic capacitor
J1—Jack to match amplifier connector
Q1, Q3—GE17 silicon general purpose audio amplifier (Radio Shack 276-2013, or equal)
Q2—GE1 germanuim general purpose audio amplifier (Motorola HEP G0002 or equal)
R1, R2, R3—15,000 ohm carbon resistor
R4, R14—2200 ohm carbon resistor
R5, R13, R16—10,000 ohm carbon resistor
R9—68,000 ohm carbon resistor
R10—1.5 megohm audio taper potentiometer
R11—500,000 ohm carbon resistor
R12—18,000 ohm carbon resistor
R15—4700 ohm carbon resistor
R17—220,000 ohm carbon resistor
R18—12,000 ohm carbon resistor
All carbon resistors 1/2 watt, 20% tolerance
S1—Single-pole, normally closed, press-to-open switch

Fig. 10-5. Circuit for a device which simulates the "swishing" sound of the surf as it breaks on shore.

A perforated board is the easiest way to hook up the unit. The values of the resistors are not all that important, so 20% tolerances are good enough. There are pins sold by electronic parts stores that fit the holes in the board. These can be used for tie points and can also serve as connection points for the transistor leads. You might want to experiment with other PNP transistors. Some might give you more "white noise".

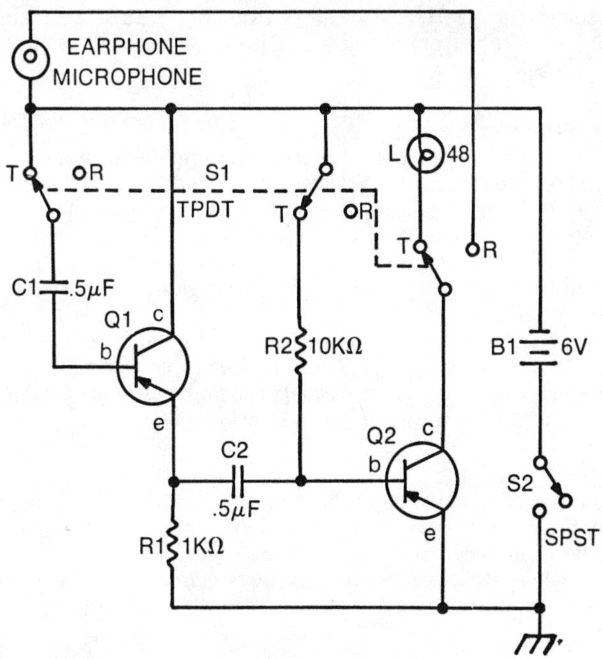

PARTS LIST
B1—6-volt lantern battery
C1, C2—0.5 μF mylar tubular capacitor
L—No. 48, 6-volt flashlight bulb
Q1—MRD-300 photo-multiplier power transistor
Q2—Any general purpose NPN power transistor
R1—1000 ohm, 1/2 watt carbon resistor, 20% tolerance
R2—10,000 ohm, 1/2 watt carbon resistor, 20% tolerance
S1—Three-pole, double-throw switch
S2—Single-pole, single-throw switch

Fig. 10-6. A photo-multiplier and one transistor permits communication with a light beam as the carrier. Unit both receives and sends.

A small amplifier should be used to drive a loudspeaker. since the signal produced in the Sound of the Surf is not powerful enough to drive speakers. However. it might work with some headphones.

TALKING OVER A LIGHT BEAM

This gadget should provide hours of fun for the experimenter and for the user. We usually think of transmitting sound through pressure waves in the air. We also know that sound can be transmitted under water. A skindiver can hear the sound of a motorboat a long way away. But. with the unit illustrated in Fig. 10-6. the voice is transmitted by means of waves in the visible region—light waves—being the medium.

Circuit Description

A light beam is projected through a lens for maximum concentration. This beam becomes the carrier for the voice and is modulated by it. The receiving unit—or the unit in the receiving mode at that moment—accepts the modulated light which now increases the current flow in the collector-emitter circuit of the photo-diode. Q1. Although Q1 is also an amplifier. further amplification is needed. This is done by Q2. in this case. although a small amplifier module could be substituted.

As you can see by examining the circuit. each unit is both a sender and a receiver. The three-pole. two-position switch (S1). converts the unit from receiver to sender. and vice versa. Sockets should be used. particularly for Q2. as you might like to try several to get the most amplification.

Construction

A plastic case will do well for housing this gadget. You will need to build two to get the effect of talking over a band of light. A lens should be placed in front of the photo-diode at exactly the distance which is equal to the focal length of the lens. This can also be done by trial and error. Avoid leaving the photo-diode in sunlight. as the lens in front of it will burn the diode.

PHASE LOCK LOOP

Every mono FM receiver is capable of being converted to receive stereo (multiplex). The 38 kHz subcarrier containing the second channel is part of the FM signal and it merely takes a circuit to separate the two signals from each other. In the past. such a circuit was complex and required several tuning coils which had to be carefully aligned to obtain the proper left and right separation. Now. with one IC and a few assorted discrete components. you achieve excellent results. Furthermore. the IC (the LM1800) eliminates all coils and the tuning is accomplished with a single pot. Even if you don't have an oscilloscope. the very minor adjustment required is done by ear with good results. Refer to Figs. 10-7 and 10-8.

The Phase Lock Loop Stereo Demodulator. as this IC is called. enables you to "pull" out the hidden subcarrier. which is present in every stereo FM program. What the IC actually does is regenerate the subcarrier. which is ordinarily (in mono operation) buried in the signal. The IC also does something which previously had to be done manually. In the older tube multiplex receivers. when you suspected a station was broadcasting in stereo. you then switched the receiver to the stereo mode. Now, the IC does that for you without waiting for your command. But, apart from switching to the multiplex mode automatically, the IC also lights up a LED so that you know you are in that mode.

This single IC provides excellent separation between the two channels. As an example:

At 100 Hz separation is 40 dB
At 1000 Hz separation is 45 dB
At 10 kHz separation is 45 dB

As frosting on the cake, the LM1800 is quite tolerant of supply voltages. It will operate equally well from 12 to 18 volts. And, the power supply can be of the simplest type since the IC has a built-in regulator. Total harmonic distortion is a low 0.5%, which is low indeed—at the recommended input level of 500 mV rms at 1000 Hz.

Circuit Description

The signal with its subcarrier enters the IC and is separated into two channels. Each channel has two discrete transistors which are used as amplifiers. Any general purpose NPN transistors can be used for that function. The LED goes on whenever the station to which the receiver is tuned is broadcasting in stereo.

A PC board is a must for this unit. If you do not want to make your own, an arrangement was made with the supplier listed in the parts list to furnish a PC board with a complete kit of parts.

Construction

The entire unit should be enclosed in a metal chassis, if it cannot be fitted within the same chassis as the receiver. In the photograph (Fig. 10-9) you can see the adapter, tuner, and power supply very clearly. The parts are spread out for visibility and should be placed closely together. However, even in this breadboarded form reception is very good.

The ON-OFF switch for the power and the knob for tuning to the stations are all on the front panel as is the LED. Power, if the voltages match, can be obtained from the receiver's own power supply. See Fig. 10-10 for a finished unit.

Alignment

Once the PLL is connected to the FM receiver, the output of the adapter is fed to a stereo amplifier connected to a pair of loudspeakers. It might require mentioning here, that the PLL is connected to the receiver just before the discriminator stage.

A scope can be used to fine tune volume control R1, which provides the degree of channel separation. Without benefit of the scope, all you have to do is choose a fairly strong station, turn the pot to one side until you lose the stereo effect and then turn back to the other direction until you lose it there. The correct balance would be between those two points at which the separation diminishes. With that done, select another station you know is also broadcasting in stereo, but which you receive weakly. Go through the same procedure as above. See if more fiddling with the pot improves the reception.

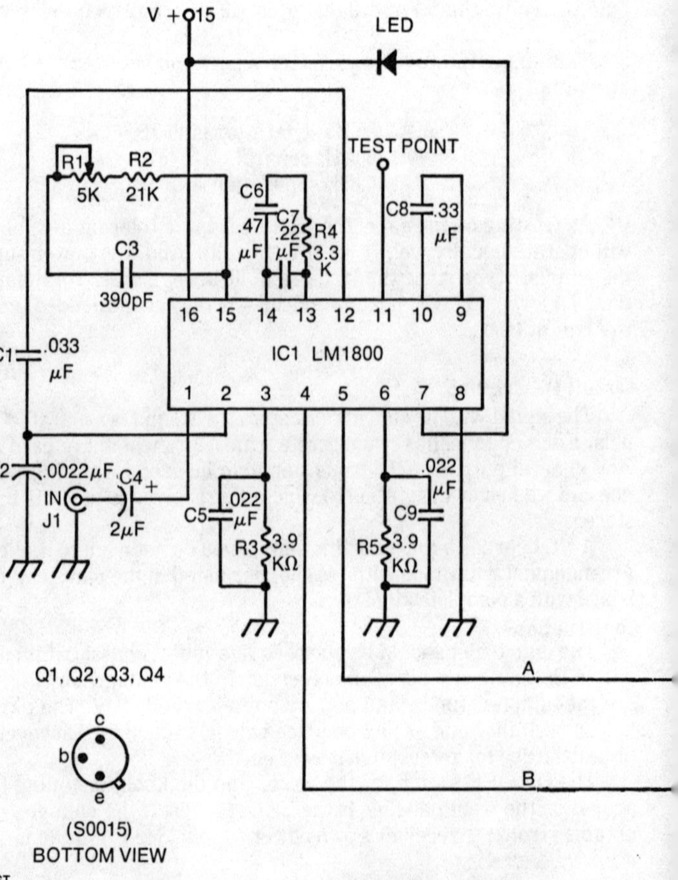

PARTS LIST

C1—0.033 μF mylar tubular capacitor
C2—0.0022 μF mylar tubular capacitor
C3—390 pF silvered mica capacitor
C4, C10, C11, C13, C14—2 μF, 25-volt electrolytic capacitor
C5, C9—.022 μF mylar tubular capacitor
C6—.47 μF mylar tubular capacitor
C7—.22 μF mylar tubular capacitor
C8—.33 μF mylar tubular capacitor
C12, C15—10 μF, 25-volt electroylic capacitor
IC1—LM1800 phase-locked loop stereo demodulator IC
J1, J2, J3—Phono jack
Q1, Q2, Q3, Q4—S0015 medium-power general purpose amplifier transistor (Radio Shack 276-2009, or equal)
R1—500 ohm potentiometer
R2—21,000 ohm carbon resistor
R3, R5—3900 ohm carbon resistor
R4—3300 ohm carbon resistor
R6, R13—1800 ohm carbon resistor
R7, R14—180,000 ohm carbon resistor

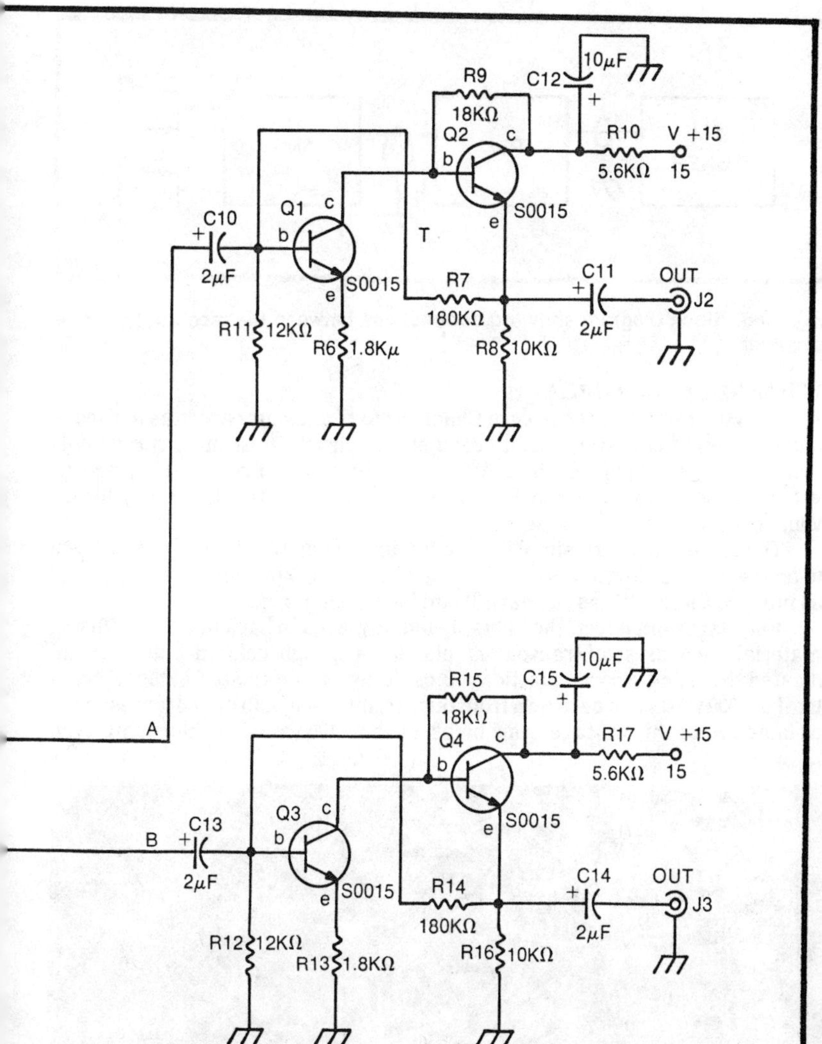

R8. R16—10,000 ohm carbon resistor
R9. R15—18,000 ohm carbon resistor
R10. R17—5600 ohm carbon resistor
R11. R12—12,000 ohm carbon resistor
All carbon resistors 1/2 watt, 10% tolerance

All parts, including PC board, but without chassis, is included in a kit available from:
RAMCO Electronic City
365 Canal Street
New York, N. Y. 10013

Fig. 10-7. Circuit for a Phase Lock Loop demodulator adding stereo to any FM receiver.

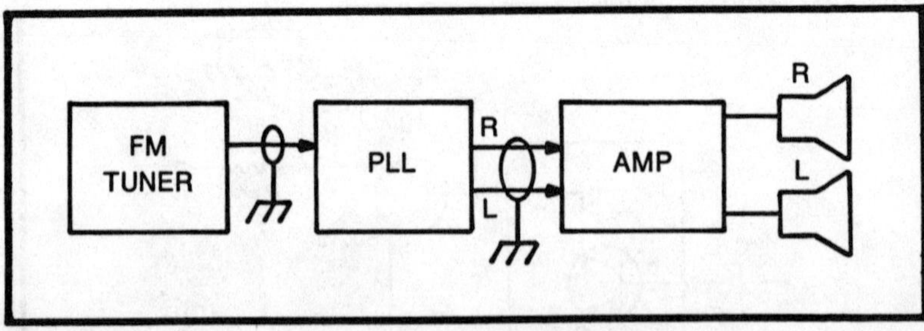

Fig. 10-8. Block diagram showing connections between the receiver, PLL, and amplifier.

3-CHANNEL COLOR ORGAN

With the microphone mixer in Chapter 8 to begin your career as a discoteque operator, the following unit based on the cirucit illustrated in Fig. 10-11 will add the finishing touches. After all, can any self-respecting disco be without flashing lights? Even if you aren't running one, you might like this for your recreation room.

The circuit is a very simple one, but with enough flexibility to permit you to choose at what levels you want the lights to go on. You can, if you wish, add as many of these units as you have 8-ohm loudspeakers.

For maximum effect, the lights should be placed in back of some diffusing material such as semi-transparent plastic. Although colored bulbs are indicated, by using colored plastic, you could use white frosted bulbs. Since a total of 1000 watts can be drawn from this circuit, each bulb may draw as much as one-third of this wattage. Nine bulbs, each at 100 watts, could be employed

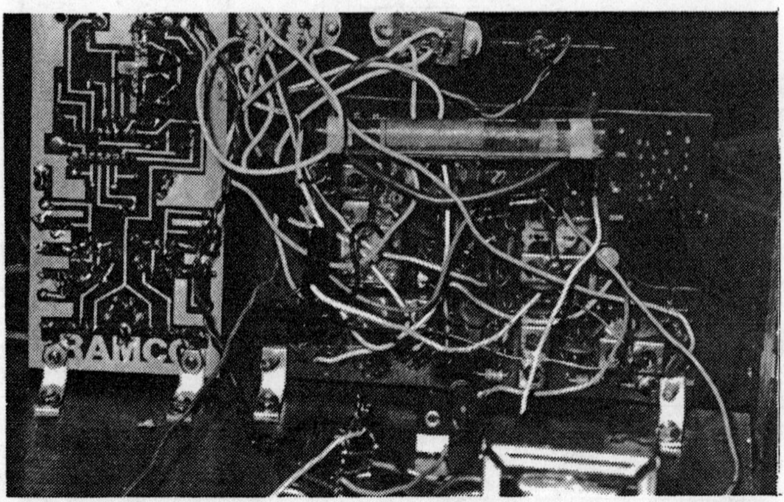

Fig. 10-9. MPX board wired (on extreme left) with FM receiver on right and power supply in front.

Fig. 10-10. Authors' own prototype of MPX wired with FM tuner. Unit is on side of recording console.

instead. There should be a margin of safety, since a cold lamp draws much more current than one that has been in operation for awhile.

Circuit Description

To achieve the desired pulsating effect from music at various levels, three silicon control rectifiers (also known as thyristors) are used for control. The circuit of Q1, consisting of R1, C1, and R4, is affected only by frequencies from 1 kHz and upwards. Music within that frequency band drives the gate of the SCR until a current level is reached at which current flows through the G and C junction. When this reaches a critical value, the gate opens and the SCR conducts. This conduction results in the bulb lighting up as long as those frequency levels are maintained. The same principle applies to each of the other two circuits; the only difference being the frequency range in which they operate. The capacitors (C1, C2, and C3) determine the frequency. The greater the capacitance, the lower the frequency to which that part of the circuit is sensitive. The circuit in which C2 is found covers the 100 Hz to 1 kHz range with C3 going from the lowest bass tones up to 100 Hz.

Construction

Since current is high in this unit, care should be observed in making certain that neither side of the AC line voltage is grounded to the case. Make doubly certain that a rubber grommet is used for the hole through which the AC wire enters.

The color organ's input is connected to the speaker by means of an audio cable. A completed unit is shown in Fig. 10-12.

WHITE NOISE GENERATOR

Does noise bother you? Are the kids screaming next door or someone else's amplifier going full blast and making it impossible for you to concen-

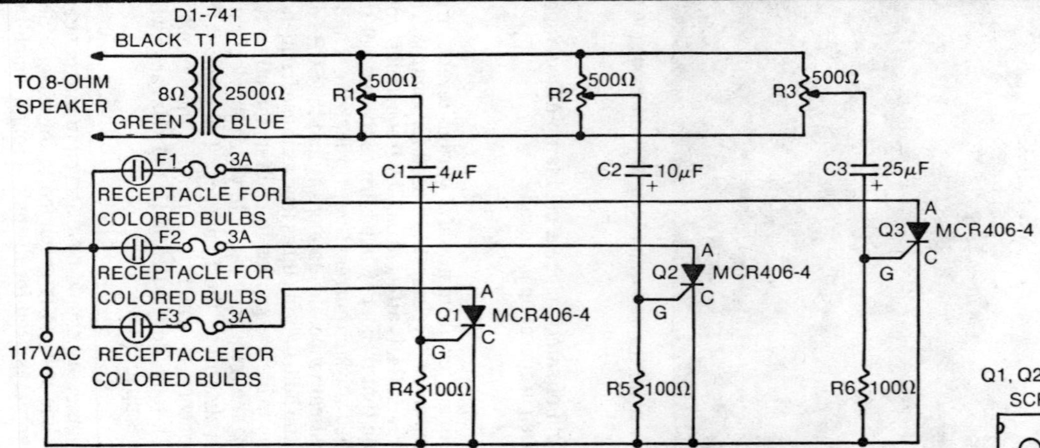

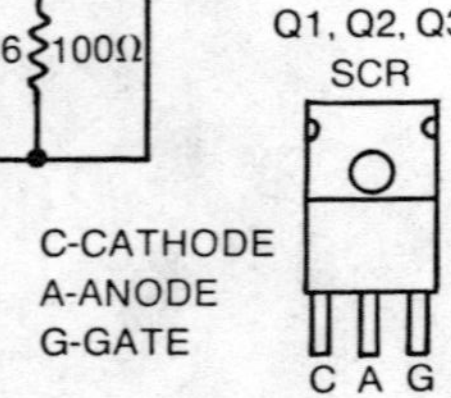

PARTS LIST
C1—4 μF. 50-volt electrolytic capacitor
C2—10 μF. 50-volt electrolytic capacitor
C3—25 μF. 50-volt electrolytic capacitor
F1 F2. F3—3 ampere. 250-volt fuse
Q1 Q2. Q3—Silicon controlled rectifier. 4 ampere. 200-volt (MCR406-4 or HEPR1218) (Radio Shack 276-1067. or equal)
R1 R2. R3—500 ohm audio taper potentiometer
R4. R5. R6—100 ohm carbon resistor. 1/2 watt. 10. . pc tolerance
T1—2500 8 ohm output transformer (GC Electronics D1-741. or equal)
Misc —Light bulb receptacles for colored bulbs. PC board. chassis 5 × 7 × 2 inches (Lafayette 12R81955. or equal)
All parts. with the exception of the chassis. is available as a kit from:
RAMCO Electronics City
365 Canal Street
New York. N Y 10013

Fig. 10-11. Circuit for a 1000-watt 3-color organ and photo of completed unit. It changes colors depending on the frequencies of the music being played.

Fig. 10-12. A completed 3-color unit.

trate? The unique circuit illustrated in Fig. 10-13 masks noise by making noise of its own. It sounds strange, but "white noise," which is the type of sound that this generator makes, seems to reduce other noises. This "white noise" is random in frequency as well as in amplitude. Its ability to cover other noise has made it useful in offices to cover the sound of people's conversations and is great in a dentist's office. The white noise masks the sound of the drill which annoys the patient even before the dentist says, "Open, please."

The white noise does not actually cancel other sounds, but due to its randomness it is ignored eventually by the ear which also ignores the other more objectionable sounds.

Circuit Description

The white noise is achieved by not connecting the collector of Q1. This converts the semiconductor to a reversed-biased diode. Reversing the base-emitter junction produces some 60 mW of random noise.

By itself, Q1 will not be sufficient to drive a loudspeaker, so RCA's CA3020, a wide band power amplifier, has been included in the circuit to raise the power output level. Since this IC requires only 35 mW to drive fully, there is enough power to spare...and with the aid of the output transformer, we change the output impedance to drive a loudspeaker.

Potentiometer R9 serves to attenuate the input signal which otherwise would result in clipping the output of the IC.

Construction

Point-to-point wiring on a perforated board is the simplest way of handling this circuit. There needs to be no exterior control for R9, since once it is set to your satisfaction it can be left alone. The unit should be enclosed in a metal chassis with the loudspeaker mounted within. The 9-volt battery also goes into the box, mounted with a battery clip.

If you were to use a larger speaker, you could mount the whole circuit inside the speaker enclosure. The current demand is low so the battery should last a long time.

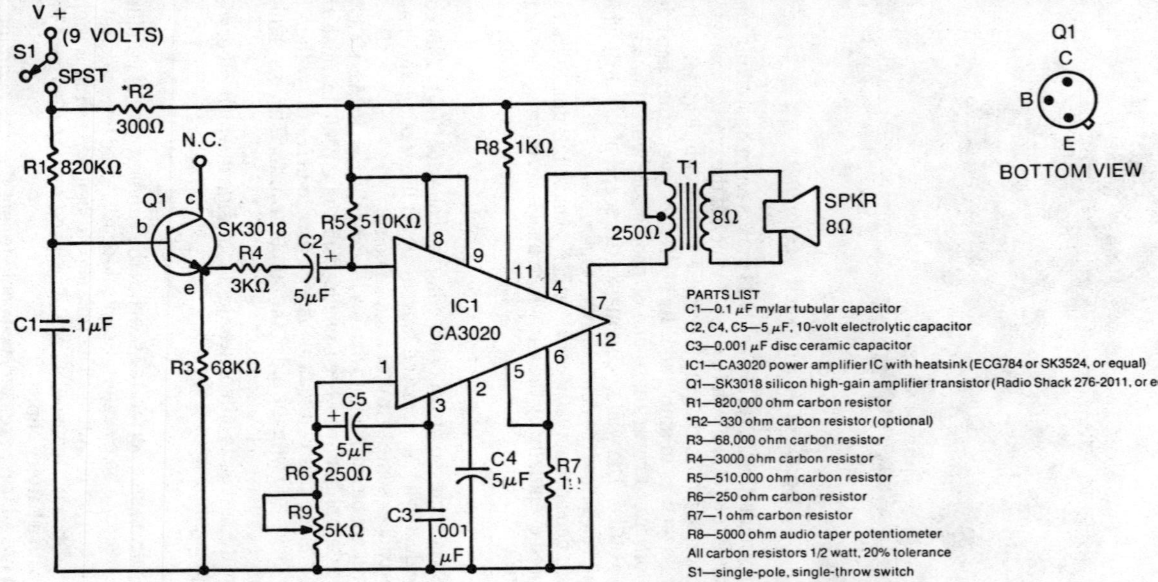

PARTS LIST
C1—0.1 μF mylar tubular capacitor
C2, C4, C5—5 μF, 10-volt electrolytic capacitor
C3—0.001 μF disc ceramic capacitor
IC1—CA3020 power amplifier IC with heatsink (ECG784 or SK3524, or equal)
Q1—SK3018 silicon high-gain amplifier transistor (Radio Shack 276-2011, or equal)
R1—820,000 ohm carbon resistor
*R2—330 ohm carbon resistor (optional)
R3—68,000 ohm carbon resistor
R4—3000 ohm carbon resistor
R5—510,000 ohm carbon resistor
R6—250 ohm carbon resistor
R7—1 ohm carbon resistor
R8—5000 ohm audio taper potentiometer
All carbon resistors 1/2 watt, 20% tolerance
S1—single-pole, single-throw switch
SPKR—8 ohm speaker
T1—250:8 ohm output transformer with center-tapped primary

Fig. 10-13. White noise generator to mask out annoying sounds. Only one IC is used in this circuit.

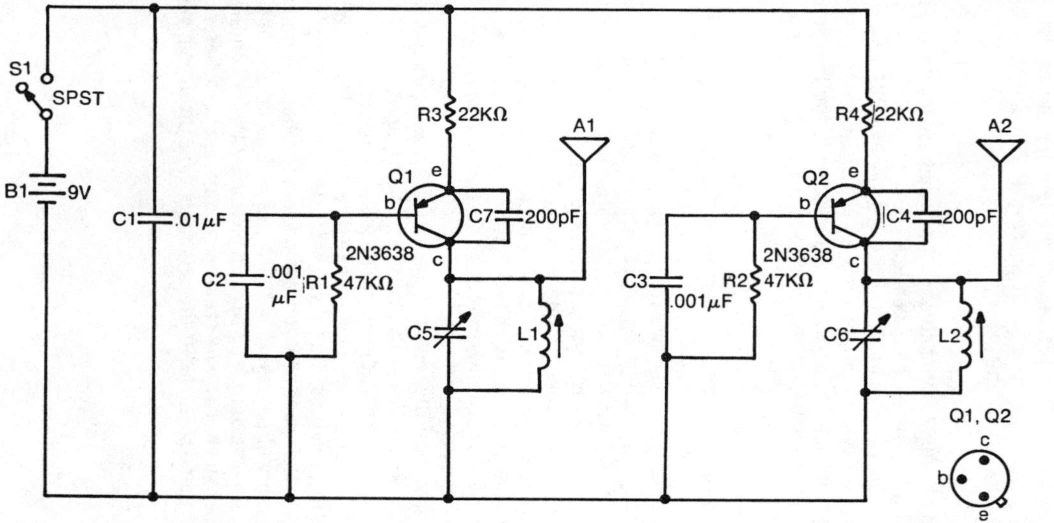

PARTS LIST

A1, A2—miniature whip antenna
B1—9-volt transistor radio battery
C1—0.01 μF mylar tubular capacitor
C2, C3—0.001 μF disc ceramic capacitor
C4, C7—200 pF silvered mica capacitor
C5, C6—200 pF variable capacitor
L1, L2—Vari-loopstick
Q1, Q2—2N3638 medium-power amplifier transistor (Radio Shack 276-2021, or equal)
R1, R2—47,000 ohm, 1/2 watt carbon resistor
R3, R4—22,000 ohm, 1/2 watt carbon resistor
S1—Single-pole, single-throw switch

Fig. 10-14. An electronic instrument of yesteryear, brought up to date using solid state circuitry. Unit produces "spaced-out" music.

PORTABLE THEREMIN

This instrument was probably the first electronic music maker. With its weird "spaced-out" sounds, it used to be the favorite background music for science fiction movies. For a while Theremin, the inventor of the instrument, used to do concerts with it, but it never caught on. Now that our ears are used to synthesized music, he could be a great success today. So, here is your chance to do what he failed to accomplish.

Circuit Description

As you can see in Fig. 10-14, both halves of the circuit are identical with Q1 and Q2 being used in a common base configuration, as are the oscillators which are tuned to the broadcast band. The radio frequencies differ by a beat-note in the audio range. A typical frequency is 1.0 MHz with a beat note of 10 kHz. This note is heard from a radio which is placed near to the Theremin. As one hand approaches the antenna, the tone varies in frequency while the other hand raises the volume.

Construction

The unit is built on a small piece of perforated board using an all metal chassis as an enclosure. The size mentioned in the parts lists houses all of the components with space to spare. The two antennas are mounted at each end of the box approximately seven inches apart. One word of caution: the transistors' symbols are drawn with the emitter rather than the collector on top, so be careful not to confuse them when wiring the unit.

Since you are dealing with radio frequencies, the leads should be as short as possible and no part of the circuit should come in contact with the box.

Four holes are drilled on top of the chassis for the two antennas and for the two loopsticks which are placed midway between the two antennas.

Place the Theremin as close to the back of an AM radio as possible. Turn on radio and find a place near the center of the dial where no station is heard. Turn on the Theremin. Place C5 and C6 at about mid-setting. A hissing sound should now be heard as you slowly adjust C5. The sound comes from the radio's loudspeaker. Adjust C6 until a piercing whistle-like tone is heard. Without changing the dial setting, readjust C5 until you obtain the lowest pitch possible.

Place your hands in the space between, but without touching the antennas. With some practice you will be able to obtain tones and go up and down the musical scale. Even if you can't play "My Old Kentucky Home" on the first try, you can always say you were really creating music from the spheres.

Why was the unit made portable? If the neighbors gang up on you, you can pick up the unit and run.

CHAPTER 11
TEST INSTRUMENTS

Why test instruments? You have followed the schematic correctly, the solder joints were done properly with not too much and not too little solder and it flowed at the right temperature. You carefully observed the hints and cautions.

You turn on the power and you fully expect the circuit to do exactly what it was designed for.

But, nothing happens, or worse yet, there is the smell of something burning! You quickly cut off the juice. First, you wonder why the circuit behaved the way it did; second, how much damage was done? Can this IC or these transistors be used, or are they destined for the junk heap?

This is the most dramatic example of the time when you need test instruments. But, there are many other times—and these are the most common—when the circuit works, but not the way it should. There is hum or noticeable distortion. How are you going to find out what is wrong? To find out the why and what requires more than a tongue-moistened finger.

You are going to need some test instruments. However, we are not talking about the laboratory type equipment that costs thousands of dollars and much training to be able to analyze the results. The hobbyist is fortunate in that most audio circuits operate on about 20% tolerance. With few exceptions, a circuit that calls for a 1 megohm resistor will operate satisfactorily with one rated at 1.2 megohms or 800K ohms. This means then, that the test instrument does not require an unusual accuracy either. Usually, the tester should be one magnitude better than the circuit being tested. Of course, the better and the more accurate the instrument, the better off you are. But, the test instruments described in this chapter have enough accuracy for the projects listed in the book; they are quite simple to build and will cost much less then any you can buy ready-made.

Our present technology demands more accurate measurements than previously, as in the days of the tubes. It would be more correct to say the measurements are of smaller quantities. A tube circuit might have demanded 240 volts. It would operate quite well with 200 volts or even 260. A tube would turn purple and could tolerate excessive higher voltages for several seconds before saying, "Goodby." A transistor does not have that leeway. Within a microsecond, an extra volt has reduced the transistor to uselessness. So, our voltmeter should be able to distinguish small differences in voltages.

Something else we demand from our test instruments...lack of distortion. In the old days audio amplifiers had 5 to 10% distortion. Some of our better present day amps have their distortion measured in hundredths of a percent. Therefore, a signal generator to be useful must have no significant distortion. Otherwise, the oscilloscope is showing the generator's distortion and not that of the amplifier.

The test instruments in this chapter are designed to meet three criteria: they must give you reasonable accuracy, be not too expensive, and be simple to build.

Which leads us to: What test instruments do you need? Which are really necessary and which would be nice to own, but you can do without? At least do without for a while.

Let's say that you now own no test instruments. Which should be first to appear on your workbench? Without hesitation, the answer would be a Volt-Ohm-Meter. It measures volts—AC and DC—resistance and DC current. Building one is not too practical a project. The switching is unusually involved. So, instead you will find a high-impedance solid-state voltmeter. It will not measure resistance, nor AC voltage (we have another project for that), but it will not load down a circuit being measured. In the material that surrounds that project, you will find out why a high impedance input is necessary, if you do not know already.

But, if you already own a VOM, how do you use it on solid state circuits which would be loaded down by the ordinary VOM? A project is included herein for you to add to your VOM so that it will be changed from a 20K ohms-per-volt to some 10 megohms per volt...a very significant difference.

The next instrument to be considered is a signal injector. There are several in the book to suit various needs. One is extremely simple. It produces a tone; if your amplifier is working correctly, the tone reproduced at the output is greatly amplified. Then there are more complex wave generators that will tell you exactly what is wrong with your circuit. With those generators an oscilloscope is recommended as a companion instrument. A scope measures voltage, as well as current and resistance; the latter two, indirectly. The scope, with its TV-like tube, shows distortion and the other gremlins that inhabit the workbench of the experimenter. You will also find sine wave generators described, one for making square waves and another for triangle waves. They each have their function.

Another instrument we recommend would be a transistor tester. You may find that a particular stage in the circuit under test is at fault. Is the transistor at fault? Was it installed incorrectly? Is one of the components that surrounds it bad? One of the projects in this chapter is designed to test transistors while they are in the circuit. Another project requires the transistor to be removed from the circuit, but it will provide more information about the semiconductor. This is why it was always advised to use sockets during construction. It avoids the time consuming, and often damaging desoldering process. However, the socket itself could be at fault. This is where a continuity tester comes in. One of the pins in the socket could be broken. A very simple project, taking only an hour or two to put together, is described for your construction.

You may have a box of unmarked zeners or diodes. You bought them cheap, but you don't know what they are. One of the instruments you can build will tell you which is which and what condition it's in.

In the appendix on resistor and capacitor codes, you will find out how to read the color codes that manufacturers use to indicate the values of their components. Resistors have one uniform code; capacitors have several.

The capacitance meter described comes in handy to give you the value at a glance regardless of the color code. Lately, some capacitors have appeared on the market devoid of any markings whatever. Unless your memory is better than most, you will wonder some days later just what is the value. The meter will remove any doubts as to what capacitance that blob of plastic happens to be.

This brings us to the most important tool of all...one that everyone owns, but that sometimes suffers from neglect. What is it?

Your brain.

Before you perform major electronic surgery on a nonworking circuit, stop and think. What are the symptoms? Act the way a surgeon does before he throws the patient on the operating table—ask questions. In this case, ask yourself. If the sick one is a stereo amplifier—are both channels dead? If so, the problem could be in the power supply. If only one channel is out of order, make comparison readings with the good channel. Do the voltages at various points agree? Within 20%? Does each channel consume the same amount of current? Are the ground connections secure? Or do they just look good? This is where the continuity tester will be of help. Take resistance readings. Sometimes a hairline crack on a PC board can drive you up the wall before you find the cause. Here, a magnifying glass can be a big help. Has solder flowed from one pin of an IC to another? Between two copper strips on the PC pattern? Look for the obvious first. Forgetting to plug in the unit is not uncommon. Sometimes a blown fuse can give you grey hair. But, don't replace the fuse before you have found the cause. The blowing of a fuse is a symptom only. Find out why it blew.

No matter how impossible something appears, check it out. Within the author's own experience was an impossible resistance reading. A resistor was color coded to indicate 100K ohms with a 1% tolerance. But, the readings came out wrong. After much scalp scratching we came back to that resistor. We unsoldered one end of it and connected an ohmmeter. It turned out to be a 10K unit—it was miscolored!

Testing is, above all, an exercise "of the little grey cells". A roomful of expensive equipment will not equal the application of judgment and logic, not overlooking patience.

We mentioned a scope some time back. It is more than a nice instrument to have. A general purpose scope is not too expensive and is a worthy addition to the workroom.

HIGH IMPEDANCE DC VOLTMETER

There is a tendency to regard a VOM as indicating some precise value of ohms, current and voltages. In the latter case, serious errors occur when measuring a low voltage, low current circuit with the usual VOM. The average priced multimeter has a 20,000 ohms-per-volt movement. We read a voltage in a solid state circuit and are fully satisfied that what we see on the scale is the voltage existing within that circuit. But in actual practice, how accurate is it? We are not writing now about accuracy of meter movement, but rather that the results of our use of a 20K ohms-per-volt meter can be extremely deceiving.

Here is an example: See Fig. 11-1. There we have a 100K resistor (R3) between a power supply and a transistor. Let us say further that a current of 100 μA is flowing through the resistor. The voltage must therefore be 10 volts. How do we arrive at that? Ohm's law says:

$$E = I \times R \text{ or } 0.0001\text{A} \times 100{,}000 \text{ ohms}$$
$$= 10 \text{ volts}$$

But, instead of figuring out the voltage, say we place our trusty (?) meter across the resistor. What we are now doing is using our meter in the 10V range, which means we are imposing a 200,000 ohm resistance in parallel with R3. So, what is the actual resistance now "seen" by that circuit? The equation for parallel resistors is:

$$Rx = \frac{R3Rm\ (\text{meter})}{R3 + Rm}$$

$$= \frac{100{,}000 \times 200{,}000}{100{,}000 + 200{,}000}$$

$$= 66.7\text{K ohms total resistance}$$

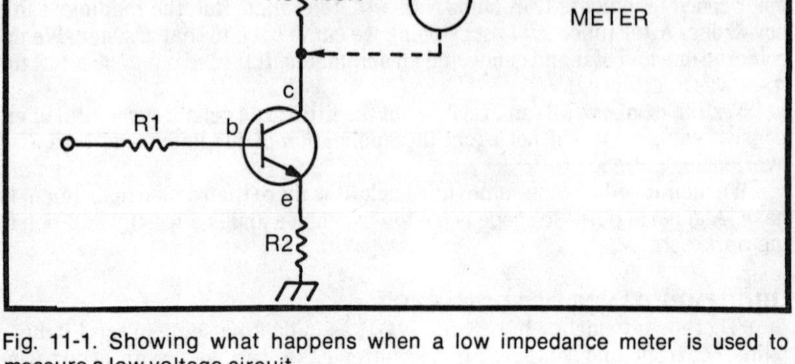

Fig. 11-1. Showing what happens when a low impedance meter is used to measure a low voltage circuit.

Now what is the actual voltage when a 100μA current is passing that new resistance?

Again we refer to Ohm's law:

$$E = IR$$
$$= 0.0001 \times 66{,}700$$
$$= 6.67V$$

Instead of the expected 10 volts (which would be correct) our meter has indicated a voltage which is 33% off. This is a serious error.

What can you do about it? You could, each time you make a measurement, calculate the percentage of error for each reading. There is fortunately, a much simpler way without discarding your present meter. The simple add-on circuit illustrated in Fig. 11-2 will convert your VOM to a high impedence unit. By raising its input impedance to 11 megohms, the loading of the circuit under test becomes negligible.

This is the advantage of the VTVM. But, as its name suggests the Vacuum Tube Volt Meter uses tubes which require some five minutes to stabilize and since the unit requires high voltages to operate the tubes, it is not portable.

The circuit in Fig. 11-2 requires no waiting period and is instantly ready to go when you turn on the power, which is only a 9-volt battery. It has been designed as a "front end" for the VOM, but if you have a reasonably accurate 50 μA meter youcould wire that in the circuit instead and free your VOM for other uses. If you would rather employ the VOM, connect its 50 μA scale into the circuit by means of the probes on the VOM.

The Field Effect Transistor (FET) has a high input impedance similar to that of a vacuum tube. At first thought, that was the obvious choice. However, experimenting brought out one serious complication. Since the circuit is essentially a bridge across which the voltage is being measured, two FETs were needed. This meant drift and the meter had to be constantly adjusted for differences in temperature. The solution was found in a dual FET—two FETs in one case—which eliminated the problem. No difference in internal temperature exists and therefore no drift.

The brains of the voltmeter is the dual FET, represented in the circuit as Q1 and Q2. Applying a DC voltage causes a rise of the source current for Q1. This now causes a difference in potential between the sources of Q1 and Q2. Connected as a voltmeter between these points, the meter indicates this balance in volts. The meter really measures the same voltage in each selected range. What we do is add resistances which total 11 megohms. Each range adds a certain amount of that resistance, so that the readings remain within the province of the meter.

Construction

If your VOM is goint to serve as the meter, then a case measuring 3.5 × 6.5 × 2 inches is adequate. If a separate meter is going to be incorporated, then a much larger case must be chosen; its actual size would depend on the dimensions of the meter. It pays to get a meter as large as possible, as it makes reading the scale that much easier.

To maintain further accuracy, close tolerance resistors should be chosen for R1 through R5. Certainly no more than 10% and 5% would be better. Current drain is low, so an ordinary 9-volt battery will last a long time.

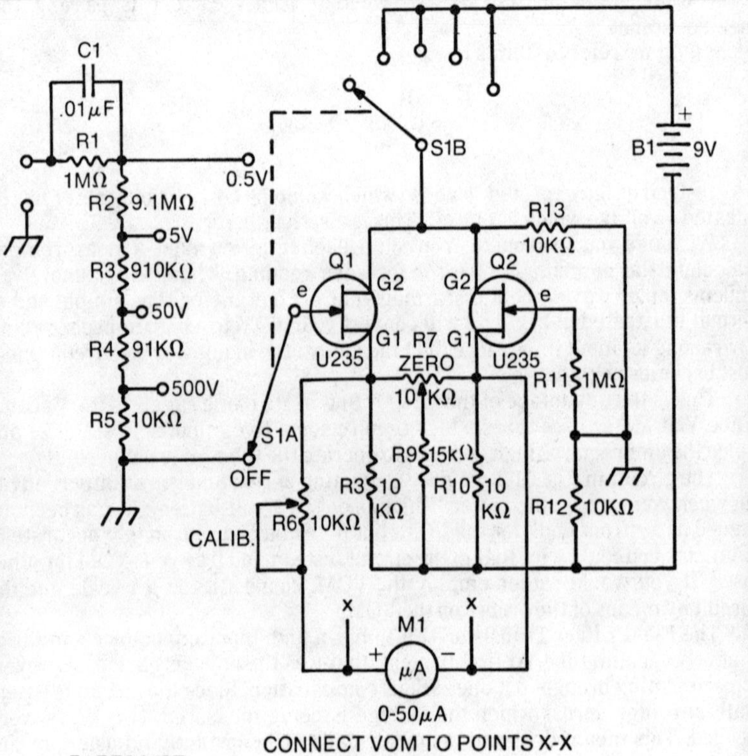

PARTS LIST

B1—9-volt transistor radio battery

C1—0.01 μF mylar tubular capacitor

M1—0-50 μA meter (optional)

Q1, Q2—Dual J-FET transistor (Siliconix U235, or equal)

R1, R11—1 megohm carbon resistor, 5% tolerance

R2—9.1 megohm carbon resistor, 5% tolerance

R3—910,000 ohm carbon resistor, 5% tolerance

R4—91,000 ohm carbon resistor, 5% tolerance

R5—10,000 ohm carbon resistor, 5% tolerance

R6, R7—10,000 ohm linear taper potentiometer

R8, R10, R12, R13—10,000 ohm carbon resistor, 20% tolerance

R9—15,000 ohm carbon resistor, 20% tolerance

All carbon resistors 1/2 watt

S1—Double-pole, 5-position switch

Fig. 11-2. Schematic for a high-impedance voltmeter with the ability to measure any DC voltage from 0-5 to 500 volts, with little or no effect on circuit being measured.

Wire switch W1 so that, when it is turned from the OFF position, it reaches the highest range first. This prevents an overvoltage for that range with the needle consequently being banged out of shape, not to mention burning out the meter.

Both of the binding posts, J1 and J2, should be insulated from the chassis. Use red and black 5-way posts. These are mounted on the case front and so should the range switch S1. The two pin jacks were placed on the top of the case and the VOM's probes were connected to them. Being put where they are, the two sets of probes did not become entangled. (See Fig. 11-3.) A small perforated board handles the few components. Trimpots should be chosen for R6 and R7. These are (respectively) for calibration and for zeroing the meter. They too, should be mounted on the board, since once they are adjusted you will not need to touch them again.

Calibration

Touch the two probes together and zero-set the meter. This is true whether you are using your VOM or a separate meter. The easiest way to calibrate the meter is to use a voltmeter that is known to be accurate and use that for a standard. If you cannot borrow one for the occasion, obtain three fresh C or D size batteries. By connecting them in series you have a 4.5-volt standard for the 5-volt range of the meter. Slowly adjust R6 until you have the correct reading.

If your range resistors are close tolerance units, the other ranges will be just as accurate as the 5-volt range. A dab of varnish or nail polish on the pot and you can close the case. Only when readings become erratic will you need to change the battery.

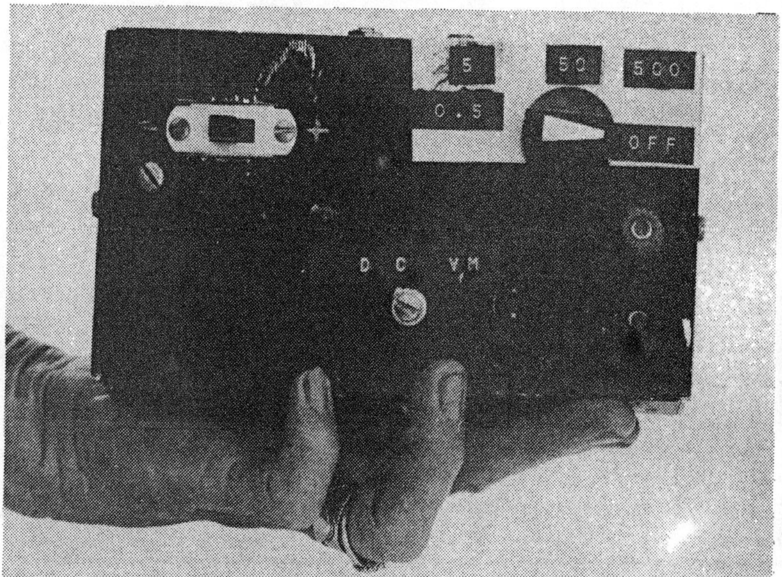

Fig. 11-3. A high-impedance voltmeter.

AC MILLIVOLTMETER

The schematic in Fig. 11-4 shows a meter circuit which will read 10 mV (0.01 volt) at full scale. The average VOM, which may have a 20K ohms-per-volt sensitivity DC, will have only half that much in the AC ranges. So, for the

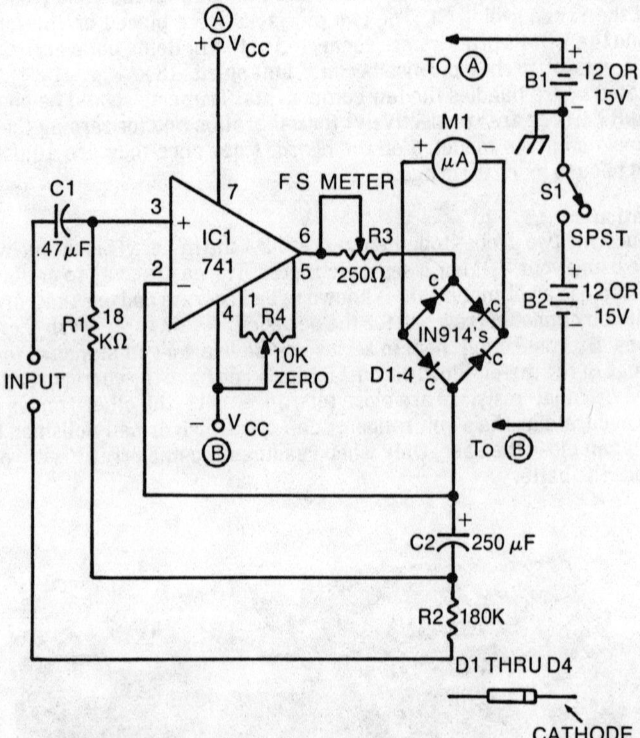

PARTS LIST
B1, B2—12 or 15-volt battery
C1U0.47 μF mylar tubular capacitor
C2—250 μF, 25-volt electrolytic capacitor
D1, D2, D3, D4—1N914 diode (Radio Shack 276-1122, or equal)
IC1—741 operational amplifier (Radio Shack 276-007, or equal)
M1—0-50 μA ammeter
R1—18,000 ohm, 1/2 watt carbon resistor, 10% tolerance
R2—180,000 ohm, 1/2 watt carbon resistor, 10% tolerance
R3—250 ohm linear taper potentiometer
R4—10,000 ohm linear taper potentiometer
S1—Single-pole, single-throw switch

Fig. 11-4. Basic design for an AC millivoltmeter using one IC. Any desired ranges can be added to the unit.

reason given in the section on the DC Voltmeter, you will load down the low voltage circuits which are prevalent these days. However, the unit described here has a 1 megohm input impedance and is accurate with signals up to 150 kHz. With such an elevated impedance, there will be very little inaccuracy when measuring a low voltage AC circuit.

Circuit Description

The heart of the millivoltmeter is an internally compensated op-amp, the 709. It is similar to the old work horse, the 741, except for the compensation. An AC signal is applied to the non-inverting input of the op-amp. This is pin 3. The current is amplified and appears at output pin 6. The current passes through the bridge and is rectified by four diodes, D1 through D4. The reading on the meter is an average value of the full-wave rectified current.

Construction

The circuit is best built on a small piece of perforated board. Use a socket, its shape depending on whether you have an op-amp in a round or in a DIP package. This socket should be glued to the board with plastic cement. The components are then soldered to the socket pins. The diodes are also set on the board, as is trimpot R3. This trimpot is used to adjust the meter to indicate full scale when a 10 mV signal is applied. This allows you to use either a 50 or 100 μA meter. As a matter of fact, you can use the same meter that you used for the DC Voltmeter, or employ the one on your VOM. Once the trimpot is set, it is not necessary to reset it.

By adding suitable attenuators, the same meter can be made to indicate full scale with greater AC signals.

Note that a positive and negative voltage supply is needed for this op-amp. Use two 12 or 15-volt batteries, or devise a power supply of your own.

BASIC DIODE TESTER

A very simple but effective circuit for testing diodes is exemplified in Fig. 11-5. Its simplicity is deceiving, as it can tell you the basic facts about an unknown diode. First, it indicates the polarity of the diode: the needle on the voltmeter moves only if the diode in question is connected with its cathode (its positive side) to the plus terminal of the meter. Reversed, there will be a sharp voltage drop. Second, if the diode is open there will be no current flow in either direction, and so, no meter movement. Third, in the case of a shorted diode, it will act as a bare wire and the full voltage will appear on the meter.

It is hardly necessary to make a permanent hook-up of this circuit. It is meant as a quick way for now-and-then testing. Use clips to make the few connections. It will take all of three minutes and will not tie up the voltmeter for other uses. For more definitive testing, the circuit which follows should be considered.

TRANSISTOR, FET AND DIODE TESTER

There are times when you want to know more about a transistor than merely Go–No Go. You need more detailed information about the device. One such time is when you must match two transistors to equalize two circuits. A tester is needed which will allow you to measure the transistor's leakage.

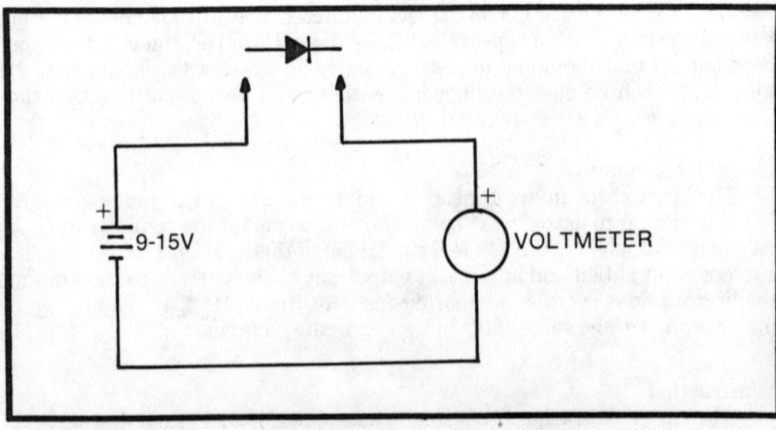

Fig. 11-5. A simple GO-NO GO diode tester.

collector-emitter current and its beta, or gain. With a FET in the tester, you obtain a measurement of its I_{DSS}, the drain to source current with the gate shorted to source. This is probably an FET's most important parameter. This parameter tells you its quality and is a help when matching is needed.

With a diode in the tester, you quickly know whether it is silicon or germanium and its condition. That's a lot to get from one simple tester. (See Fig. 11-6.)

Circuit Description

Three pin jacks are labelled E-D, B-G, and C-S. In these you respectively insert the emitter of a transistor or the drain of a FET, the base or the gate, and the collector or source. Diodes are inserted in the C-S and E-D terminals. Polarity is unimportant. With the reversing switch (S3) in one position a diode will give a full scale reading or else it's dead. In the other position, the meter will denote the reverse current, which should be a low reading. Note that there are two ranges and, even in the 1-mA position of the range switch (S1), the reading will still be low.

Whether transistors or FETS are being tested, their polarities can be reversed without harm to the devices. If an FET is inserted incorrectly, pressing S2 will not indicate a drop on the meter as it should. It should be remembered that FETs are symmetrical in construction. The gate is half way between the drain and the source. So, it is possible to interchange the drain and source leads with no change in the operation of the FET within a circuit. But physically, the lead of the drain is not between the leads of the source and the gate. So, the pin connections of that particular device should be observed very carefully before thrusting the FET into a circuit. The current will be read on the 10 mA scale. Depressing switch S2 connects the gate so that it is reversed biased with respect to the source. I_{DSS} is typically 0.1 to 10 mA.

With a transistor inserted in the jacks, and the polarity switch S3 in the correct position, the current will be read on the 1 mA scale. This places R4 in series with the meter. Depressing Gain Switch S2 indicates the beta of the transistor.

Construciton

Any wiring method can be used successfully. Perf board with terminal pins is as good a method as any. Always wire the switches before placing them in their positions on the chassis face. The size of the chassis depends on the size of the meter you are going to use. Any surplus meter will do; preferably it should have an internal resistance of 50 ohms. If it doesn't, the shunt resistor, R4, should be calculated to suit.

All three pin jacks should be insulated from the case. The 1 mA position of S1 should be labeled T and the 10 mA labeled F.

ZENER DIODE TESTER

In every hobbyist's junk box are unmarked zeners which makes them almost useless for any practical application. As a matter of fact, lots of "top hat" rectifiers look like zeners and it's nice to know which is which. One way is

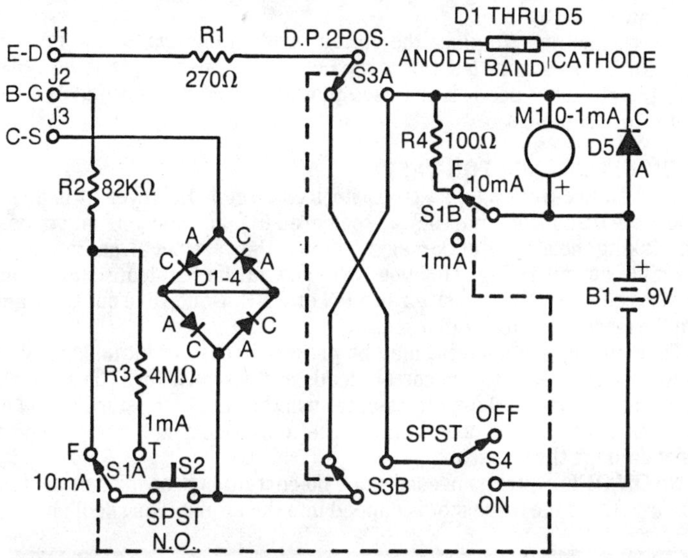

PARTS LIST

B1—9-volt transistor radio battery

D1, D2, D3, D4, D5—600-volt (PIV), 1 ampers silicon rectifier diode (Radio Shack 276-1104, or equal)

J1, J2, J3—Insulated tip jack

M1—0.1 mA ammeter

R1—270 ohm, 1/2 watt carbon resistor, 10% tolerance

R2—82,000 ohm, 1/2 watt carbon resistor, 10% tolerance

R3—4 megohm, 1/2 watt carbon resistor, 10% tolerance

R4—100 ohm, 1/2 watt carbon resistor, 10% tolerance

S1, S3—Double-pole, 2-position toggle or slide switch

S2—Single-pole, single-throw momentary ON switch

S4—Single-pole, single-throw toggle or slide switch

Fig. 11-6. A tester to determine whether a transistor, FET, or diode is operative.

to use an oscilloscope which will indicate the zener's breakdown voltage. This limiting action is typical of a zener and separates instantly the zener from the ordinary diode. If you don't have a scope or want to set it up for just one zener it is much simpler to use the circuit illustrated in Fig. 11-7. It eliminates the scope, using a VOM instead. The voltmeter section of the VOM (or a separate voltmeter) is set to read AC voltages which must go high enough to reach the zener's breakdown voltage.

The circuit makes use of the AC variable supply. Resistor R1 can be anything from 200K to 300K, rated at 1 watt. The resistor serves as a current limiter. The cathode of the diode is connected to the plus side of the voltmeter. The voltage from the power supply is turned up very slowly. The meter will indicate the rising voltage until a point is reached at which the needle stops. That is the rated breakdown voltage of the zener. If you get no reading at all, the zener is either shorted or has been connected backwards. If the zener is open, on the other hand, it will have no limiting action and the voltmeter will continue to indicate the full AC voltage. If open or shorted, the zener should be thrown away.

There is no need to place the circuit inside a chassis, as the few components can be breadboarded whenever necessary. Also, there is no reason to solder the circuit, since it is not used that often and it takes only a couple of minutes to assemble it.

IN-CIRCUIT TRANSISTOR TESTER

Say you are working on a transistorized circuit. However, it is dead. You suspect the transistors, but they are soldered in place and you hate to unsolder them, taking the chance of damaging them if they are not already. With the ultra-simple circuit in Fig. 11-8, you will know of the bad transistor at once. It will also tell you if the transistor is NPN or PNP. Don't take out the transistor from the circuit, but test it in place.

Turn off any voltage that may be present in the circuit under test. Then, clip the tester's leads to the correct leads of the transistor. The schematic of the circuit you are working on indicates whether the transistor is NPN or not. But, even if you don't know that information...it does not matter. The tester cannot damage the transistor.

No ON-OFF switch is needed since no current is drawn when the tester is not in use. Once the transistor is clipped into the circuit, you should hear a tone

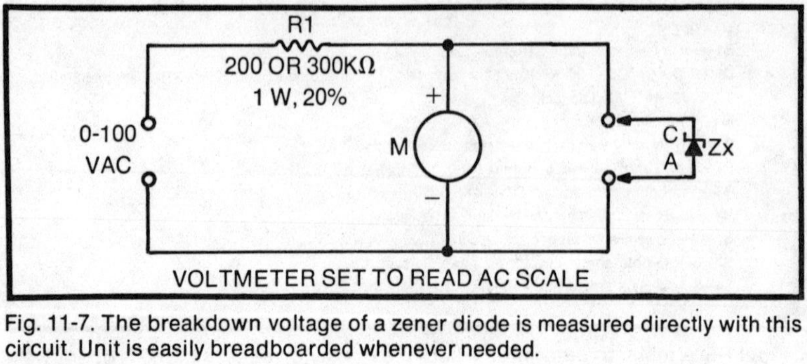

Fig. 11-7. The breakdown voltage of a zener diode is measured directly with this circuit. Unit is easily breadboarded whenever needed.

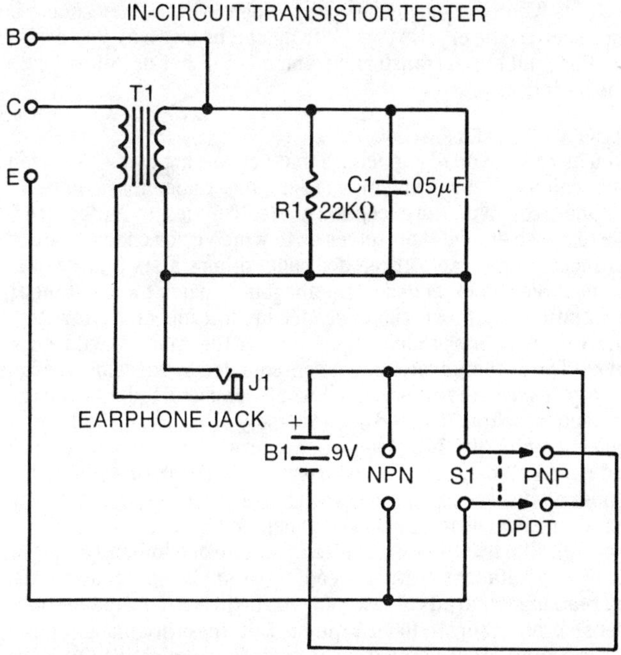

CRYSTAL OR ANY HIGH IMPEDANCE EARPHONE.
SMALL CLIPS TERMINATING EACH LEAD FROM B, C, AND E.

PARTS LIST
B1—9-volt transistor radio battery
C1—0.05 μF mylar tubular capacitor
J1—Mini-jack to match earphone
R1—22,000 ohm, 1/2 watt carbon resistor, 20% tolerance
S1—Double-pole, double-throw slide switch
T1—20,000:1000 ohm output transformer
Misc.—Crystal or any high impedance earphone, small clips terminating each lead from B, C, and E.

Fig. 11-8. The tester eliminates need to unsolder transistors from a circuit before testing. Sound of the tone in the earphone indicates quality of semiconductor.

in the earphone. If no tone is heard, move S1 to its other position. If silence still reigns, then the transistor is not operating. Usually the louder the tone the greater is the gain of the transistor. Most defective transistors can be detected in this fashion.

What you have done is made an oscillating circuit with the transistor and it is the oscillations which you hear in the phones.

Construction

Use a metal chassis and bring out the leads correctly marked. Make them short and use tiny alligator clips at their ends. S1 should preferably be a slide switch with its two positions marked as PNP and NPN. There should also be a jack which fits whatever earphones you happen to own. They should be of high impedance such as the crystal type. Wiring can be any way you prefer. R1 and C1 can be fastened to the transformer and a small battery clip should be used to hold the battery in place.

IC CAPACITANCE METER

Do you have a boxful of capacitors that you are not using because you don't know their values? Your problem is not unique. Capacitors come in a variety of shapes and sizes. With the exception of electrolytics, manufacturers seem to be unable to agree among themselves as to which color code to indicate values and tolerances. Some use a three dot code, others a six and even a nine-dot code. So you have a color coded capacitor, but you don't know if the first dot is the first significant number, the tolerance or, if it meets military specs. Then too, some tiny capacitors have appeared on the market with no markings whatsoever. These complications are compounded by the wide tolerance found in capacitors. There are times when it would be useful to know the exact value of that capacitor, so that it would not adversely affect the circuit.

Requiring only two ICs, easy to procure and inexpensive, the circuit described here will measure values from 1 pF (0.000001 μF) up to 10 μF. This covers most of the range you are apt to use. Electrolytics are usually well marked and rarely need to be checked except for leakage.

The circuit illustrated in Fig. 11-9 is a simple one to build and offers direct readings of capacitances without needing to apply an equation afterwards. What you read in picofarads or microfarads is the actual capacitance. The biggest expense is an accurate 100 μA meter. But, these are found regularly in the pages of most mail order catalogs. The case can be plastic with an aluminum cover. These are standard with GC Electronics, Radio Shack, etc. The case size depends on the size of the meter movement.

Circuit Description

The circuit is simplicity itself. IC1 is a dual Schmitt trigger, both halves being used. IC2, a monostable multivibrator, is triggered by an oscillator at preset regular intervals. The result is a square wave at a frequency with a pulse rate determined by the unknown capacitor. By clamping these pulses to a set potential, the output becomes a square wave oscillating between negative and the reference potential. The average voltage, measured by the current meter, is directly proportional to the pulse and therefore to the capacitor in question.

The experimenter may recognize that IC2 could be a 555 wired to act in that fashion. However, it would require different values for that IC. It would be worthwhile to check the chapter of this book devoted to the 555 used as a multivibrator to see how to use it.

C1 through C5, the standards, should be close tolerance units and have low leakage. The accuracy of your circuit is dependent on these capacitors and your meter.

Since the meter draws over 40 mA, the switch chosen for the ON-OFF switch is a momentary ON. The unknown capacitor is connected to the terminals, J1 and J2. Although 5-way binding posts can be used, it is more practical to use pin jacks. Solder short leads with pin plugs at one end of each

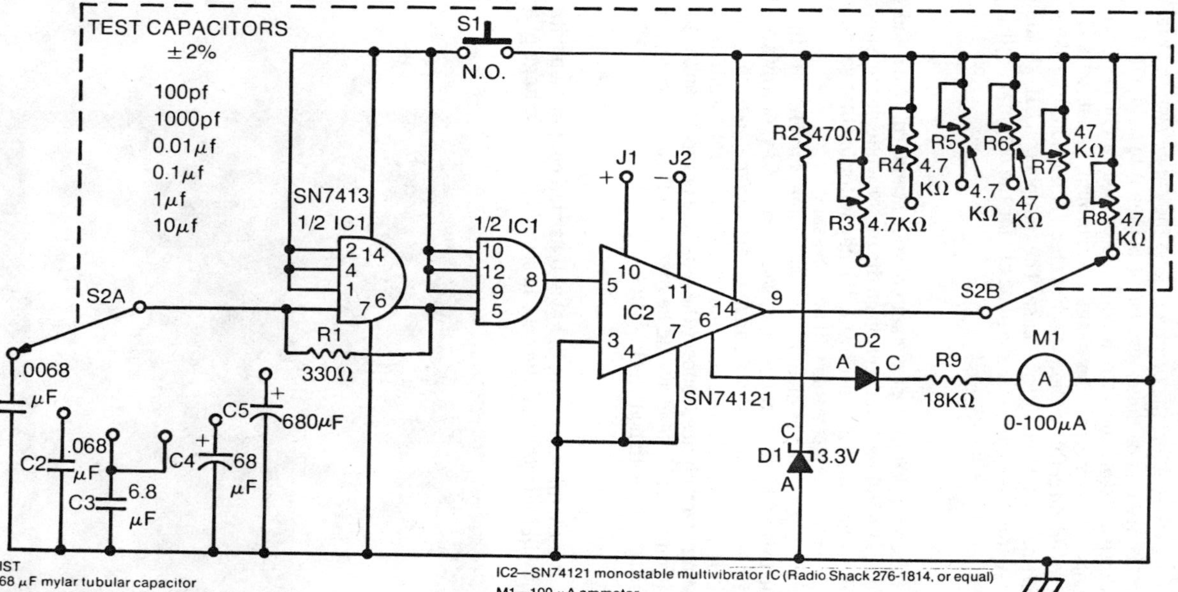

PARTS LIST
C1—0.0068 μF mylar tubular capacitor
C2—0.068 μF mylar tubular capacitor
C3—6.8 μF tantalum capacitor
C4—68 μF tantalum capacitor
C5—680 μF electrolytic capacitor
D1—1N5226B, 3.3-volt, 500 milliwatt zener diode (Motorola HEP Z0206, or equal)
D2—50-volt (PIV), 1 ampere silicon rectifier (Radio Shack 276-1101, or equal)
IC1—SN7413 dual Schmitt trigger IC (Radio Shack 276-1815, or equal)
IC2—SN74121 monostable multivibrator IC (Radio Shack 276-1814, or equal)
M1—100 μA ammeter
R1—330 ohm, 1/2 watt carbon resistor, 10% tolerance
R2—470 ohm, 1/2 watt carbon resistor, 10% tolerance
R3, R4, R5—4700 ohm linear taper potentiometer
R6, R7, R8—47,000 ohm linear taper potentiometer
R9—18,000 ohm, 1/2 watt carbon resistor, 10% tolerance

Fig. 11-9. Only two ICs are used in this circuit to read the capacitance of unknown capacitors with a high degree of accuracy.

and alligator clips at the other ends. Avoid exchanging leads of another length as it may change the readings of the meter. Short leads are recommended.

The front panel holds the meter, the terminals and the range switch (Fig. 11-10). The perforated board can be quite small as it is meant to hold only the sockets for the ICs, three resistors, plus two diodes. The board can be fastened to the screw terminals of the meter. The range determining capacitors, C1 through C5, can be wired directly to the switch wafer. This should be done carefully before inserting the switch into the case. The trimpots on S1B should be wired in such a way that they can be reached easily for the original adjustments.

Calibration

Connect a close tolerance 100 pF test capacitor, your standard, to J1 and J2. Adjust R8 until you get a full scale deflection on the meter while depressing S2. Do the same for the other ranges, substituting the capacitor for that range. Once the pots are set, no more adjustments will be necessary.

The way to determine the value of an unknown capacitor is to clip it to the alligator leads, turn to the highest range and work your way downward, depressing S2 each time, until you reach a range where the meter needle indicates a value. This is the capacitance of the unknown. No transposition is necessary. When testing electrolytics, connect the plus side of the capacitor to the lead from point 10 on IC2. This lead should be marked with a PLUS on the panel. Polarity can be ignored with non-electrolytics.

DECIBEL METER

A decibel meter is a useful instrument to have. If used in conjunction with a signal generator, it will indicate the difference in sound level or voltage of a

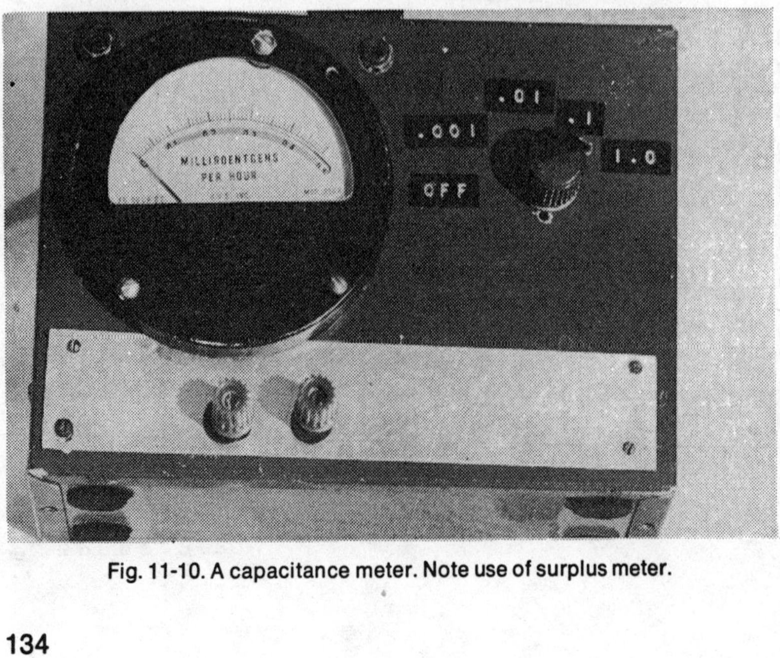

Fig. 11-10. A capacitance meter. Note use of surplus meter.

pure sine wave. This allows you to discover the amount of amplification from one stage to another of an amplifier. You can find out how flat your amplifier response is really by plotting a graph based on frequency versus level.

The unit in Fig. 11-11 can also be used as an AC voltmeter when connected across a load, such as a loudspeaker or a resistor equal in value to the speaker. The output voltage is then read on one of the three scales on the meter face. The four AC ranges are 0-1 volt, 0-3, 0-10 and 0-30. The meter scale is also calibrated in dB (decibel) with a zero reference level of 1 mW at 600 ohms.

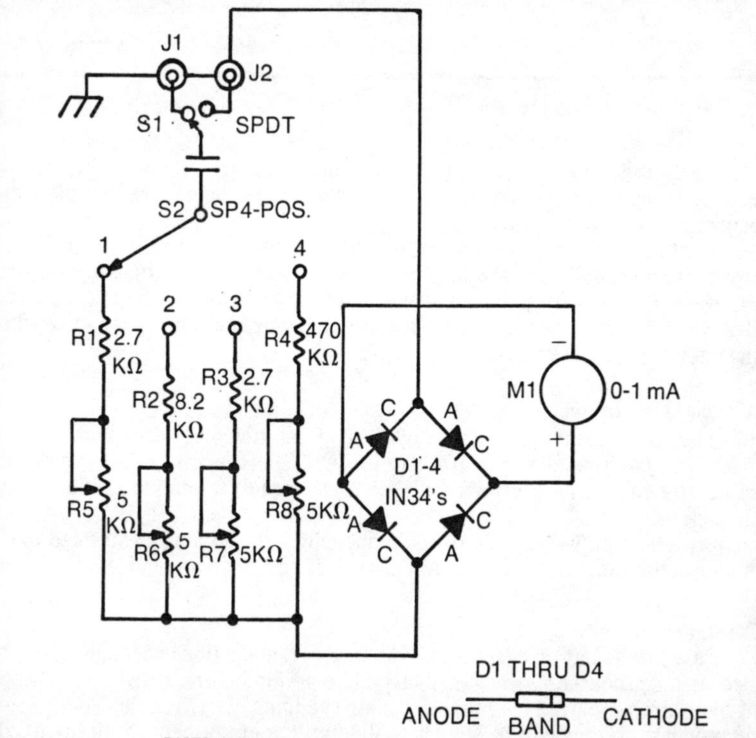

PARTS LIST

C1—10 μF, 25-volt, non-polarized electrolytic capacitor

D1, D2, D3, D4—1N34A diode (Radio Shack 276-1102, or equal)

J1, J2—Phono jack

M1—0-1 mA ammeter

R1, R3—2700 ohm, 1/2 watt carbon resistor, 10% tolerance

R2—8200 ohm, 1/2 watt carbon resistor, 10% tolerance

R4—470,000 ohm, 1/2 watt carbon resistor, 10% tolerance

R5, R6, R7, R8—5000 ohm, 5 watt wirewound potentiometer

S1—Single-pole, double-throw switch

S2—Single-pole, four-position rotary switch

Fig. 11-11. Meter to measure (in decibels) the relative output levels of amplifiers. Circuit is easy to build, requiring only a few components.

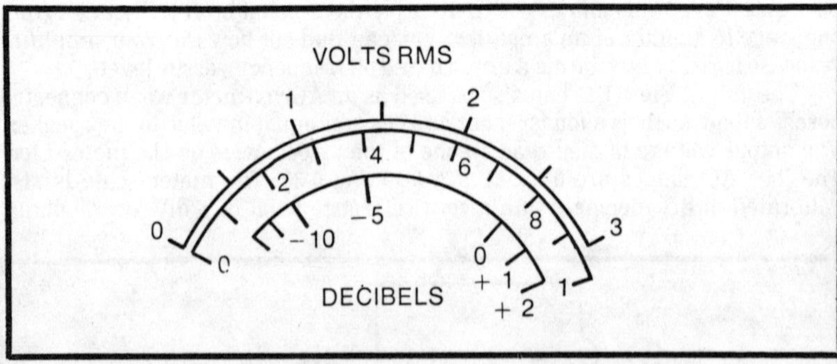

Fig. 11-12. Design for the scale to be inserted into the meter face.

This is an industry standard and corresponds to 0.775 volt. Note the scale in Fig. 11-12, which should be copied and pasted on the meter face to replace the original 0-1 mA markings.

Try to get a good meter movement that is linear and from which the glass front can be easily removed. Several electronics supply houses list such meters in their catalogs. But, in general even those not expressly designed for that purpose can be changed. It requires delicacy and avoidance of bending the needle which is generally very fragile.

Circuit Description

The input signal is attenuated by the combination of fixed and variable resistors for each range. A group of four diodes forms the bridge. The meter then reads the unbalance caused by the incoming signal, which is rectified by the diodes.

Switch S1 is designed so both channels of a stereo amp can be fed to the meter at the same time. Of course, only one channel can be read at a time.

Construction

Case size is left up to the individual builder, since this is dependent on the size meter procured. Wirewound variable potentiometers, rated at five watts or over, are employed to obtain full scale readings in each scale. They should be mounted so they can be set when aligning the instrument. Note the markings for S2 on the panel front (Fig. 11-13). A small piece of board with terminal

Markings for S2		
Volts		dB
30	–	+ 30
10	–	+ 20
3	–	+ 10
1	–	+ 0

Fig. 11-13. Panel markings for range switch.

pins is fine for mounting the components for this project. No special precautions need be taken. Wiring is straight forward.

Calibration

An accurate voltmeter is necessary for calibration of the dB meter. Connect the sine wave generator to the AC voltmeter and your instrument.

Set the output of the sine wave generator to 10 volts, having set range switch S2, to the 0-10 V range. Turn pot R5 so that the meter reads exactly 10 volts. Now reduce the generator's output to 5 volts. It should match pretty closely the center mark of the new dial. The pot may have to be readjusted slightly, as your meter might not be completely linear.

Reduce the signal generator's output to three volts. Move S2 to the 0-3 volt range. This is position 2 on the switch. Repeat the procedure as in the previous range and do the same for the other ranges.

The decibel marks will fall into place. Note that 0 dB is equal to 0.77 volt on the 0-1 volt scale.

The settings obtained will remain fixed for a long time, as the only things which can affect the readings will be a bent meter needle and the aging of the resistors, which takes a long time with proper use. Just make certain that you always begin with the highest range and work downward until you find a range in which you obtain a reading.

CHAPTER 12
SIGNAL GENERATORS

Signal generators are test instruments. However, they deserve a special chapter devoted only to them due to their versatility and applications.

They are all pretty much used in the same way—from the simplest signal injector to the multi-range, multi-wave-form generators with attenuators and such.

The experimenter, who may never have done a great deal of troubleshooting, finds himself looking at an apparently disordered maze of criss-crossing wires. He feels, at first, that he cannot hope to correct whatever has gone wrong with that circuit. However, in most cases, it is not all that difficult.

The circuit itself is not the place to start. The exception is when a wire is dangling and you can see where it should be connected. The place to begin is with the schematic. Break it up, in your mind, in "bite-size" pieces; each done with the definition of its function.

Start at the beginning. Where does the signall originate? At a jack? Locate that jack on the panel. Where does that signal go and what happens to it? In your mind trace it along the schematic so you understand the purpose of each section of the circuit. The smaller you reduce the section, the easier to understand.

For example, let us take a 3-stage audio amplifier. Each stage is meant to amplify the input signal. There are two ways to work. Each is equally good and which is better is a matter of opinion.

One way is to start at the input and work your way forward. Or you can start at the output and work backward. It's like entering a circle, one place is as good as another providing you continue to the logical end.

With our audio amplifier example, we find that by injecting a tone in the last stage, that tone comes out in the loudspeaker connected to the unit. There is little or no amplification, but at least we know that stage is not at fault. One third of the work is already done.

We move back to the second stage. No tone. This then is the troublesome one. This is where a voltmeter comes in handy, as does an ohmmeter to isolate the specific faulty component. But this is beyond the range of this text.

The second tracing method is working forward on a non-working circuit. A tone generator is clipped into the input of the amplifier. A small monitor amplifier with a tiny loudspeaker, or even a pair of high impedance headphones, is clipped into the first stage. Does it work? If it does, clip the monitor to the second stage. If your monitor has a decibel meter circuit incorporated within it, you can actually measure the amount of amplification arising out of each stage.

Whichever method is used, forward or backward, the results should be the same.

The next questions that are bound to be asked are, "Why are different waveforms used?", "Why sine, square, or triangle waves?", "Isn't a one-tone signal generator enough, why do I need all the other gadgets?"

Space will not permit going into every detail of this subject, fascinating as it is. But, here are some of the highlights. A one-frequency signal injector operates basically as a continuity tester. It tells you that there are no breaks, or if there are, where they exist. It will indicate that your amplifier can pass a 1 kHz tone, or whatever the test tone is. But, what will it do to a 400 Hz tone, or with frequencies running up to 20 kHz?

A sine wave is an AC wave. It does not have the harmonics of a square wave. The square wave generator producing a 1 kHz tone will also be generating harmonics all the way up to 10 kHz.

Without going into all of the arguments pro and con on the use of a triangle wave, which is relatively new in testing audio equipment, smaller percentages of distortion become visible. Much smaller percentages than are obvious with a sine or even a square wave. All of this is dependent on the use of a scope to be able to examine the waveform. The answer to that is, "Any audio equipment generates some distortion." So why be so fussy?

Before going any further in this argument let us end it here. You now have some idea of how to use your signal generator. The generators described in this chapter will serve your purposes. Enjoy building them and using them.

CONTINUITY CHECKER

Most continuity checkers, such as the low range on an ohmmeter, depend on your being able to see the meter face. With some, a tiny light goes on when the circuit is uninterrupted. But, they all demand that you take your eyes off your hands while holding the probes to see the results. But, what happens when you are up on a ladder checking a loudspeaker on the wall?

The continuity checker described below will indicate an unbroken line by a tone heard through its loudspeaker, which is self-contained. The whole unit is so small that it can be clipped on one's belt, freeing the hands and not requiring the use of the eyes. Refer to Fig. 12-1.

Circuit Description

Q1 and Q2 are wired as an oscillator, the output of Q2 being fed back to the input of Q1. As long as the probes are not touching or completing an unbroken circuit, the checker sees an opening, as though a switch were left open. Once

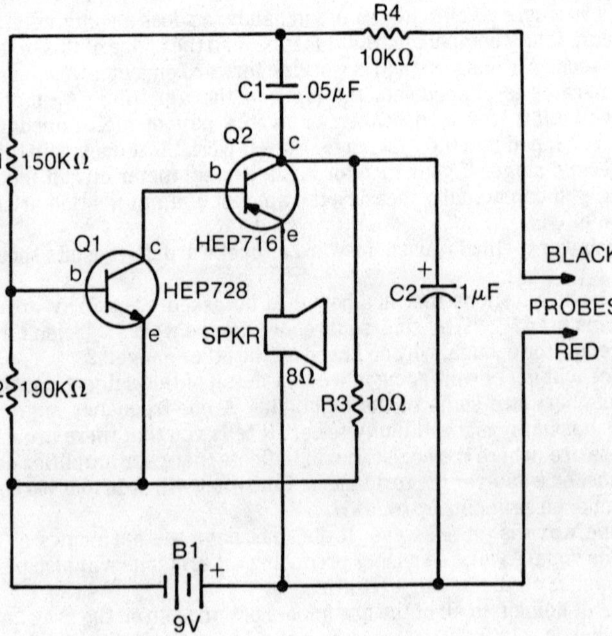

PARTS LIST
B1—9-volt transistor radio battery
C1—0.05 μF mylar tubular capacitor
C2—1 μF, 15-volt electrolytic capacitor
Q1—HEP728 general purpose silicon audio amplifier transistor (Radio Shack 276-2016, or equal)
Q2—HEP716 general purpose medium current amplifier transistor (Radio Shack 276-2034, or equal)
R1—150,000 ohm, 1/4 watt carbon resistor, 10% tolerance
R2—190,000 ohm, 1/4 watt carbon resistor, 10% tolerance
R3—10 ohm, 1/4 watt carbon resistor, 10% tolerance
R4—10,000 ohm, 1/4 watt carbon resistor, 10% tolerance
SPKR—8 ohm speaker, 1 or 2-inch diameter

Fig. 12-1. Audible continuity tester. Small enough to hook onto one's belt.

the circuit is completed, power is applied to the transistors which causes them to oscillate. R4 is a current limiting resistor.

Construction

A small plastic box, not much larger than a cigarette pack, is big enough to hold the few components. Small holes are made in the cover above the place where the loudspeaker goes. The four resistors—quarter-watt ratings are sufficient—and the two capacitors (get the smallest possible)—are mounted on a small piece of perforated board using point to point wiring.

The sockets for the two transistors are glued to the board. A battery clip holds the 9-volt battery in place. A clip should be epoxied on to the box so it can be slipped over the belt or on a breast pocket. The lead from R4 should be terminated with a black pin jack and the one from the plus side of the battery with a red jack. Various transistors have been found to work, provided that Q1 is an NPN and Q2 is a PNP. If you have several transistors you want to try, you might want to breadboard the circuit first. It will take only minutes. No switch is necessary since there is no current consumption while it is idle.

To test the unit, place the two probes so they touch. A loud tone should be heard. If not, check the polarity of the capacitor and the transistors.

SINGLE IC SIGNAL INJECTOR

If you have never done any troubleshooting, or need a small and extremely portable signal injector, the circuit in Fig. 12-2 should fit the bill. It will work with audio or even radio frequencies. Designed around the popular 7400 IC, the entire unit can be built into the probe itself. By injecting the signal into the last stage of an audio amplifier or a radio receiver and working backwards, you should hear the tone gradually increasing in volume. If you have an oscilloscope, you can also see the wave form enlarging and can observe where and if appreciable distortion is being introduced. Without the scope, you have to depend on your ears. They will at least tell you when the signal stops.

The 7400 IC injector is far from complex. Four discrete components, two AA cells, a switch, and a tiny chassis (or the probe) in which to put the whole thing and you are in business.

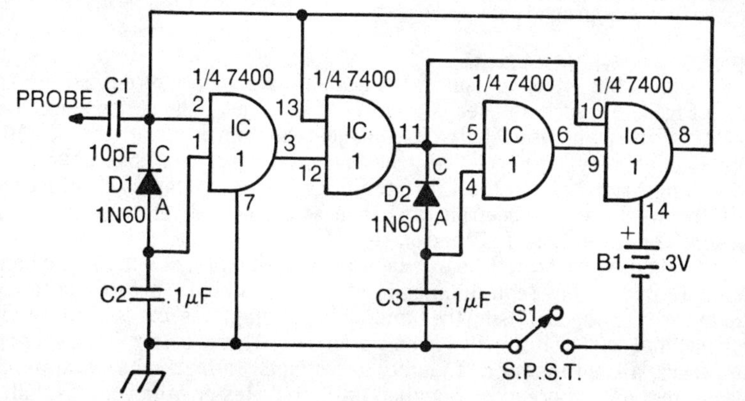

PARTS LIST

B1—3-volt battery (two AA cells in series)

C1—10 pF silvered mica capacitor

C2, C3—0.1 μF mylar tubular capacitor

D1, D2—1N60 diode rectifier (Radio Shack 276-1102, or equal)

IC1—Single-pole, single-throw switch

Misc.—Chassis, shielded cable, probe

Fig. 12-2. Four sections of the one IC are employed to make this signal injector. Only two AA cells are needed to furnish the power.

The 7400 is a 4-gate logic NAND circuit. The gates are tied together resulting in a multivibrator. Due to the short rise and fall times of the square wave generated. the harmonics of the basic frequency—which in this case is about 3 kHz—go all the way into radio and even UHF ranges. This means you could troubleshoot a TV set. But. unless you are experienced. you can do more harm than good to the set and to yourself. The high voltage inside a TV set is lethal. If you are at all careless.......

Construction

The easiest way to build the tester is to buy a probe. They are often available in electronic stores or you can improvise one from a plastic tube. With the probe in hand. make a small piece of perforated board to fit inside the probe. On the board. glue the socket for the IC and solder the components directly to the protruding pins of the socket before you press in the IC. C1 should be tantalum and C2 and C3 are ceramic discs. Make certain the pin numbering of the IC you are using is identical to the schematic. Since you are dealing with radio frequencies. shield the probe by wrapping foil around the perf board. Be careful that the foil does not touch any part of the circuit.

If you are unable to obtain a probe case large enough to contain the entire circuit. there is an alternative. Build the circuit inside a metal chassis. From the chassis run shielded cable to the probe. In this case. it can be made from an old plastic pen case. The metal point of the old pen should be soldered to the positive lead with the ground terminating with an alligator clip. The clip can be fastened to any negative point in the circuit or even to the chassis being tested. if it is at ground potential.

SINE WAVE GENERATOR

"Simple but effective" sums up the sine wave generator described here. (See Fig. 12-3.) Such a tester is of prime necessity for the serious electronic hobbyist. Its quality is such that it is adequate for any use...with the exception of a laboratory type project. The generator produces a clean sine wave with a total range from 30 Hz to 30 kHz in three steps (which overlap). This range covers the whole audio spectrum and then some. Completely solid-state. the generator uses only one FET and one IC.

Tube type generators are excellent in their stability. at least in the better units. However. they require time after turn-on to stabilize before one can use them with confidence. Using the familiar 709 op-amp. this unit is always ready for instant use. The distorition is only about 4%. This is better than necessary to service all but the highest quality audio units. In fact. many commercial generators offer more distortion than this simple tester. And yours is built at much lower cost.

It should be remembered that before you look at a sine wave—or any test signal—that has been run through an amplifier. you should first examine the wave directly on the scope without the amplifier in the circuit. Figure 12-4 shows the setup to accomplish this. The switch allows you to compare the wave before and after amplification. If you have a dual trace scope. the switch is not needed. What you are doing is memorizing the wave so that you can recognize the distortion for what it is. You now know whether the distortion is inherent in the amplifier or is in the tester.

Since the distortion in this generator comes mostly from the range-determining capacitors. C1 through C6. choose mylar for the lower values and tantulum for the electrolytics. Low tolerance resistors are also a must. Above all, do not use capacitors with more than 10% tolerance. Use a capacitance meter to choose those that come closest to the desired values. If you don't own such a meter, you might build the one described in this chapter and build it for now and for future use.

Even with 10% tolerance capacitors, you will get a lower percentage of distortion than that due to the canceling effects of one capacitor with another.

If you plan to do your work only on the work bench and portability is not a must, the generator can be operated from a well filtered, plus and minus 15 VDC power supply. For field work the unit is used instead with two 9-volt batteries. Whichever system you use to power the unit, it is ready to go as soon as it is turned on.

Circuit Description

Stability in this generator is maintained by the FET instead of the customary lamp. A lamp has two disadvantages: it is a current consuming device and it requires time to stabilize. The FET, on the other hand, requires no stabilizing time.

This is how it works: an increase in the output of the IC means a greater negative voltage appearing at the gate of the FET. Therefore, its effective resistance increases and so does the amount of negative feedback to the IC. The actual gain of the IC is unimportant in this application. Since the gain of the op-amp is in inverse ratio to the resistance of the FET, that decrease of gain counterbalances the previous gain. This results in excellent stability and no drift.

There are two feedback loops: one positive and the other negative. The positive loop goes to the Wien bridge, existing between pin 5 and pin 10. Pin 5 is the non-inverting input of the 709. This results in severe attenuation, although oscillation is maintained. The negative feedback loop offers enough gain to make up for the previous loss. The result is unity, a gain of one. The compensation network is made up of C9 and R10 between pins 12 and 3.

It was found that gains of different FETs varied considerably, so it was decided to employ trimpot R12 to set the output at the desired level.

Three frequency ranges are offered:

Position 1 of S2	30 Hz to 300 Hz
Position 2 of S2	300 to 3 kHz
Position 3 of S2	3 kHz to 30 kHz

The fine frequency control (dual pot R4 and R5) permits continuously variable output from one range to another.

Construction

A solid aluminum box as an enclosure is needed to avoid body capacitance effects which could affect the oscillation of the unit. A chassis 572 inches is about right. Holes should be drilled on the back of the chassis which now becomes the front panel. The following controls are mounted on the chassis: Coarse Frequency, Fine Frequency, and Coarse Output Level. In its first position, the Output Level Control serves as an ON-OFF switch. Before mounting the controls onto the chassis, wire the fixed capacitors and resistors that belong there. Make certain that each wafer is wired correctly in the right or-

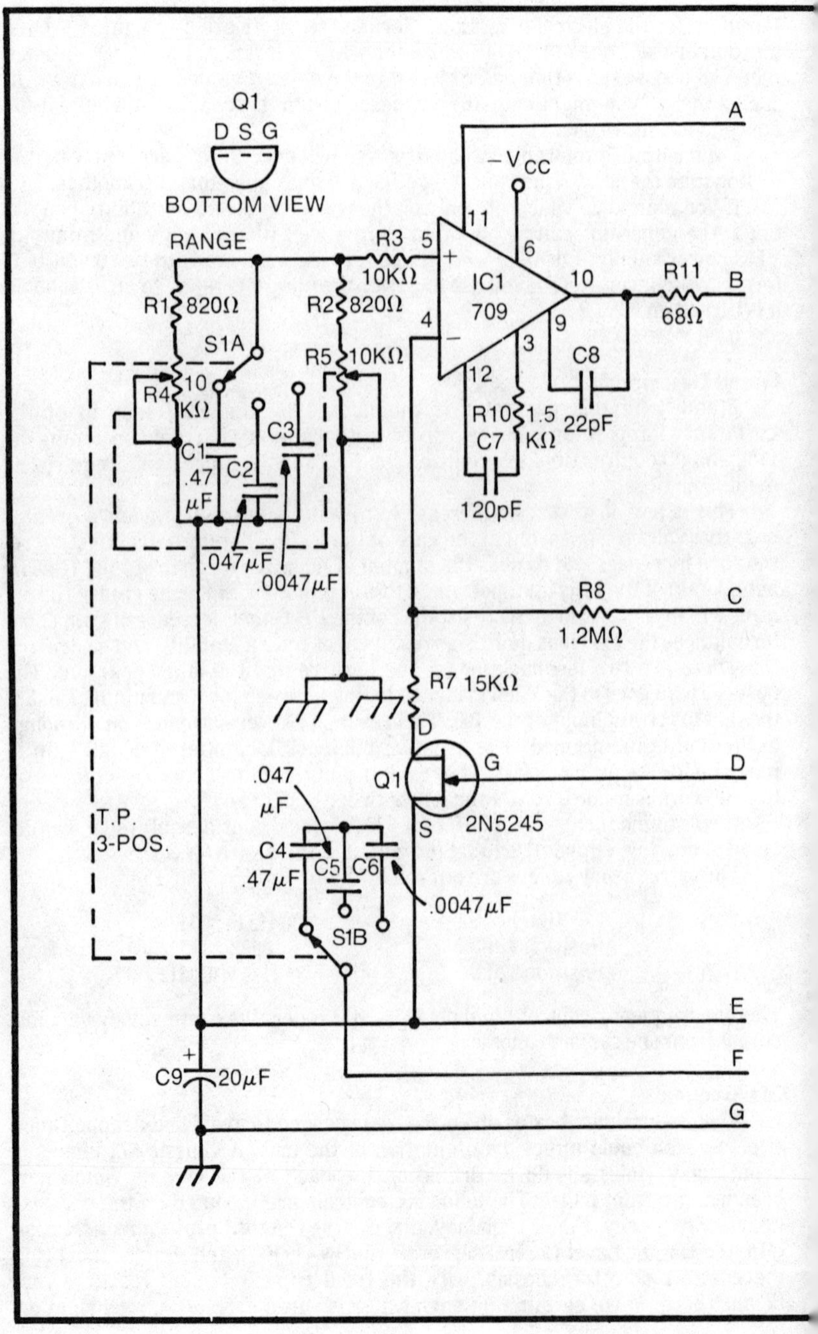
Q1
D S G
BOTTOM VIEW
RANGE
R1 820Ω
S1A
R4 10 KΩ
C1 .47 μF
C2
C3
.047μF
.0047μF
R2 820Ω
R5 10KΩ
R3 10KΩ
5
+
4
−
−VCC
11
6
IC1 709
10
9
3
12
R10 1.5 KΩ
C7 120pF
C8 22pF
R11 68Ω
A
B
C
R8 1.2MΩ
R7 15KΩ
D
G
Q1
S
2N5245
.047 μF
C4 .47μF
C5
C6
.0047μF
S1B
T.P. 3-POS.
E
F
G
C9 20μF
+

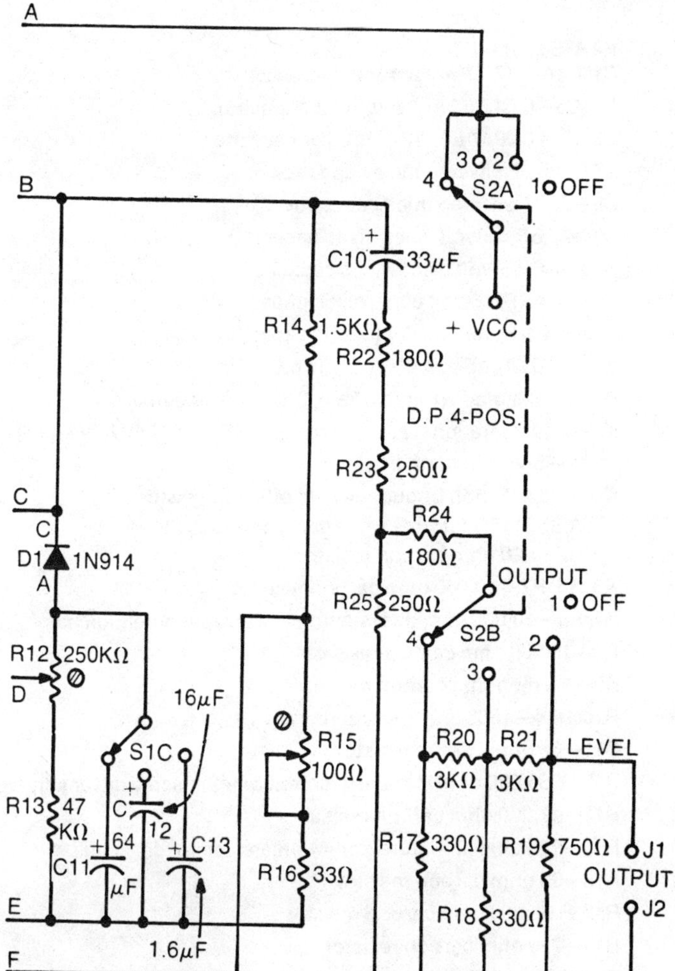

Fig. 12-3. The 709 IC is the brains of this sine wave generator. The attenuator circuit provides several output levels for various types of circuits.

PARTS LIST

C1, C4—0.47 μF mylar tubular capacitor

C2, C5—0.047 μF mylar tubular capacitor

C3, C6—0.0047 μF mylar tubular capacitor

C7—120 pF silvered mica capacitor

C8—22 pF silvered mica capacitor

C9—20 μF, 15-volt electrolytic capacitor

C10—μF, 15-volt electrolytic capacitor

C11—64 μF, 15-volt electrolytic capacitor

C12—16 μF, 15-volt electrolytic capacitor

C13—1.6 μF, 15-volt electrolytic capacitor

D1—1N914 diode (Radio Shack 276-2009, or equal)

IC1—709 operational amplifier (Radio Shack 276-017, or equal)

J1, J2—5-way binding post

Q1—2N5245 high frequency field effect transistor (RCA SK3116, Motorola HEPF0021, or equal)

R1, R2—820 ohm carbon resistor

R3, R6, R7—10,000 ohm carbon resistor

R4, R5—10,000 ohm dual section linear taper potentiometer

R7—15,000 ohm carbon resistor

R8—1.2 megohm carbon resistor

R10, R14—1500 ohm carbon resistor

R11—68 ohm carbon resistor

R12—250,000 ohm miniature potentiometer, screwdriver adjust

R13—47,000 ohm carbon resistor

R15—100 ohm miniature potentiometer, screwdriver adjust

R16—33 ohm carbon resistor

R17, R18, 330 ohm carbon resistor

R19—750 ohm carbon resistor

R20, R21—3000 ohm carbon resistor

R22, R24—180 ohm carbon resistor

R23, R25—250 ohm carbon resistor

All carbon resistors 1/4 watt, 5% tolerance

S1—Triple-pole, three-position rotary switch

S2—Double-pole, four-position rotary switch

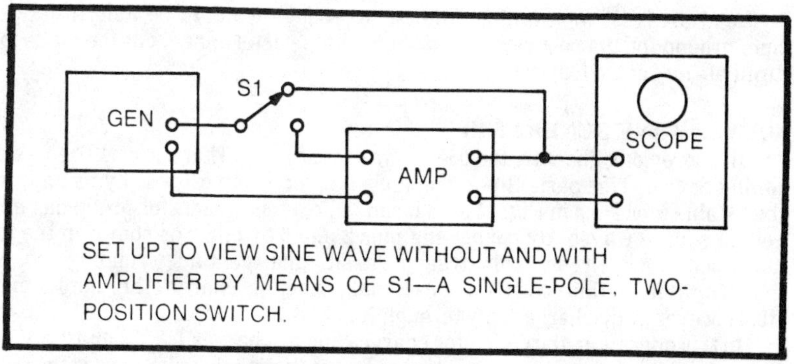

Fig. 12-4. Block diagram showing the connections between a signal generator and the amplifier being tested.

der. It is a good idea to label each wire from the switches as to where the other end is to go. The few extra minutes spent in putting a piece of paper tape with some identification is more than made up after you have been interrupted and you pick up your work another time. This avoids wondering where those different wires are supposed to go. A perforated board, with point to point wiring, is the way to do it—with terminal pins for connecting points. Sockets are recommended for the IC as well as for the FET. After the board is wires, it is then fastened to the chassis with separators to keep any of the connections from grounding. Only one ground connection should be made: this is output jack, J2, which is connected to the chassis and to the ground point of the circuit.

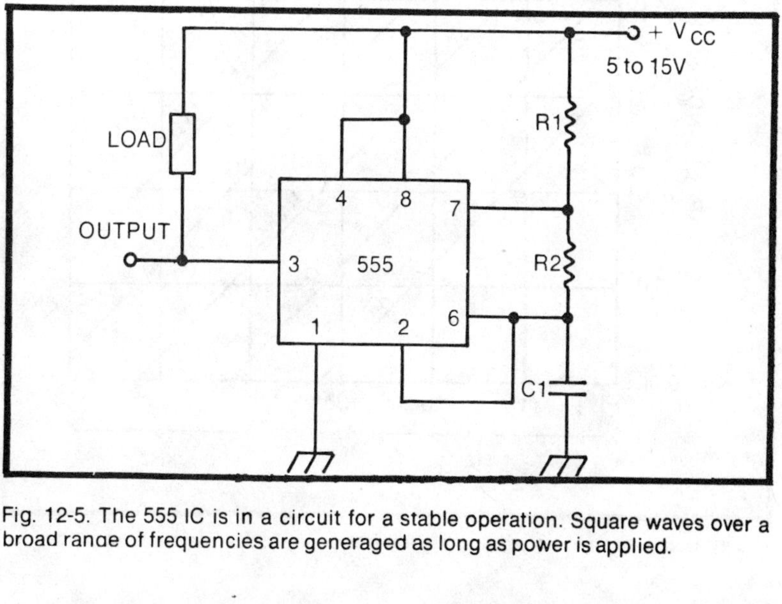

Fig. 12-5. The 555 IC is in a circuit for a stable operation. Square waves over a broad ranqe of frequencies are generaged as long as power is applied.

Lead dress is important so allow long enough leads to be able to move them around for the best wave form. Once that is determined, cut the leads to eliminate any extra lengths.

SQUARE WAVE GENERATOR

In a previous chapter, the 555 IC was mentioned. There it was used as a timing device. This basic flip-flop module has many other uses; by using it in the astable mode (as in Fig. 12-5) you can build a tone generator producing excellent square waves. By connecting pins 2 and 6 together as shown in the illustration, the circuit will trigger itself and continue running as a multivibrator. In this case C1, the external capacitor, charges through R1 and R2. However, it discharges only through R2.

In this operation, the capacitor charges and discharges 1/3 V_{cc} and 2/3 V_{cc}. Charge and discharge times are independent of the supply voltage as is the frequency. By means of the graph in Fig. 12-6, the free running frequency can be easily found.

The output of the circuit is fine for signal tracing, as the square wave shows little distortion. You may be satisfied with only one frequency. Let use say you want a 1 kHz tone. Then, R1 and R2 could be combined together so they total 1 megohm. Always remember that R1 should never be less than 1K. By following the 1K frequency line on the graph, capacitor C1 should be a 1 μF electrolytic, preferably tantalum.

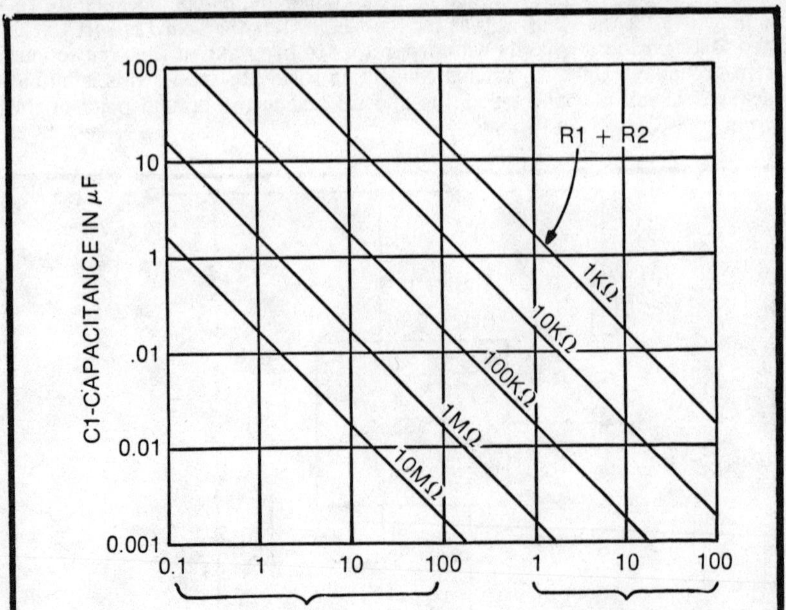

Fig. 12-6. Graph to help calculate the frequency in the free-running mode.

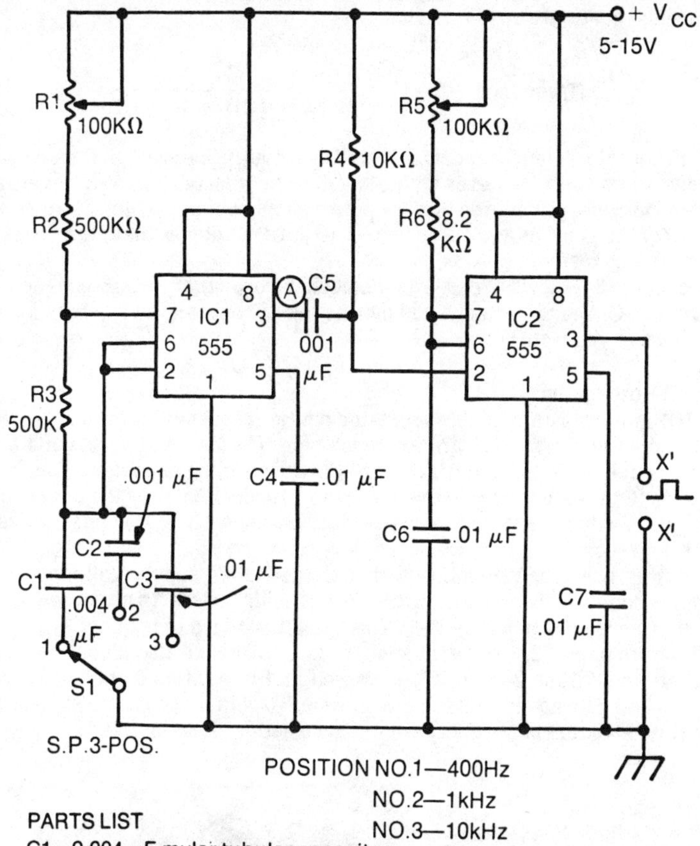

PARTS LIST

C1—0.004 μF mylar tubular capacitor

C2—0.001 μF mylar tubular capacitor

C3, C4, C6, C7—0.01 μF mylar tubular capacitor

IC1, IC2—NE555 IC timer (Radio Shack 276-1723, or equal)

J1, J2—5-way binding post

R1, R5—100,000 ohm linear taper potentiometer

R2, R3—500,000 ohm, 1/4 watt carbon resistor, 5% tolerance

R4—10,000 ohm, 1/4 watt carbon resistor, 5% tolerance.

R6—8200 ohm, 1/4 watt carbon resistor, 5% tolerance

S1—Single-pole, three-position rotary switch

Misc.—Chassis 6×4×2 inches, battery holder, test leads and probes

Fig. 12-7. A more complex square wave generator with varying duty cycles is provided by this circuit. It utilizes two 555 ICs or the dual version (556) to alter the wave form.

To calculate other frequencies not available from the graph, use the following equation:

$$f\ (\text{frequency in Hertz}) = \frac{1.44}{[2.2K + 2(Rx + Ry)]C1}$$

Frequency is independent of the voltage, but the intensity of the tone is not. So, use more voltage—never above 15 volts—for a louder tone. However, some care should be exercised so that the power supply voltage which appears at the output of the generator is not so high as to burn out diodes that might exist in a circuit being tested.

Figure 12-7 shows a more complex generator in that three frequencies are available. Of course, you can add more. Make S1 a rotary switch with as many positions as you want frequencies.

Circuit Description

IC1 is used as an astable oscillator whose frequency is controlled by S1. Make sure that C1, 2, and 3 are the mylar type. The output at pin 3, point A, is a square wave. This is fed to IC2 through C5, the coupling capacitor. This in turn triggers IC2, which is used in the monostable mode. R5 varies the duty cycle of the square wave and it can be shaped just as you wish it. For this purpose, it requires a scope.

It was previously mentioned that if the power supply voltage was too strong it could cause problems in the circuit being tested. An attenuator circuit to cure this is shown in Fig. 12-9. This circuit is added to points X – X. In the first position of S2, no attenuation occurs...100% of the signal is passed through. In position 2, only 10% is passed and in position 3 only 1%. The resistors should be no more than 5% tolerance. To obtain exact proportions of the signal, attenuators of 1% tolerance would be better.

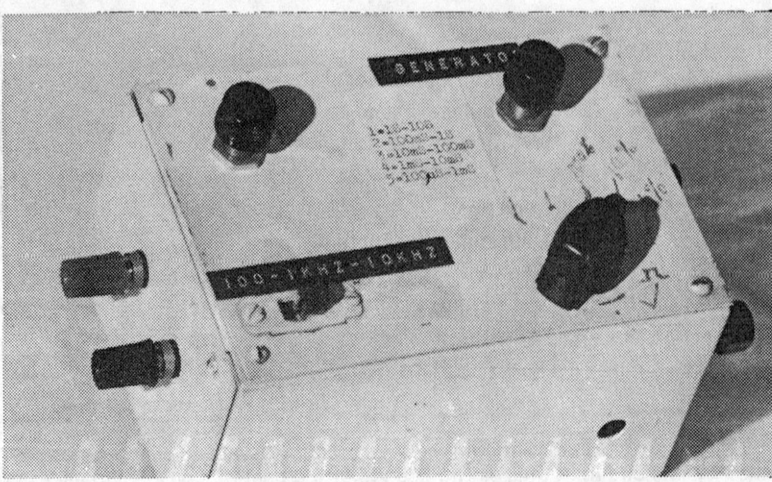

Fig. 12-8. Prototype of the square wave generator in Fig. 12-7.

Construction

The wiring requires no special precautions. For housing, a plastic box will do and an outboard power supply that provides a voltage in the 5 to 15-volt range can be used instead of building a separate supply for the generator. A small piece of perforated board will hold the few components.

On the front panel, mount Fine Frequency Control R1, wave shaper R5, and the attenuator circuit. A slide or rotary switch for Coarse Frequency

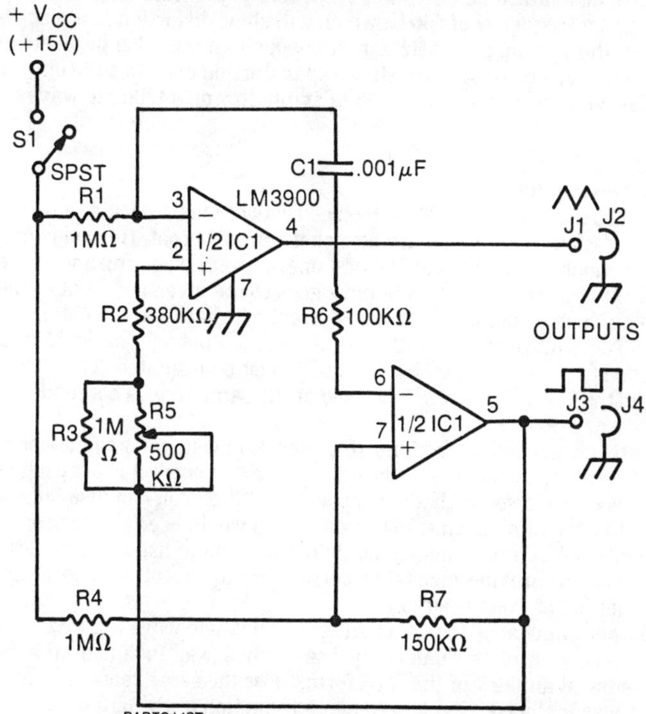

PARTS LIST

B1—15-volt battery
C1—0.001 µF disc ceramic capacitor
IC1—LM3900
J1, J3—Red 5-way binding post
J2, J4—Black 5-way binding post
R1, R3, R4—1 megohm, 1/2 watt carbon resistor, 5% tolerance
R2—380,000 ohm, 1/2 watt carbon resistor, 5% tolerance
R5—500,000 ohm potentiometer
R6—100,000 ohm, 1/2 watt carbon resistor, 5% tolerance
R7—150,000 ohm, 1/2 watt carbon resistor, 5% tolerance
S1—Single-pole, single-throw switch
Misc.—Chassis 5 × 2 × 3 inches, test leads with probes and clips

Fig. 12-9. This unit produces square and triangle wave forms simultaneously. Only one IC is needed in this low distortion generator.

Selection completes the parts on the front. The two-way binding posts can go on one side of the unit. See Fig. 12-8.

DUAL-WAVEFORM GENERATOR

Distortion is often unrecognizable when a sine wave is shown on a scope. To the eyes, it may appear perfect...even with one or two percent of distortion. With severe clipping, it appears as a square wave, which can be very misleading. Not that the sine wave does not have its uses. It does, but the unit shown in Fig. 12-9 should also be on your workbench. It provides a good stable square wave with a frequency of 5 kHz, which will show distortion in a piece of equipment all the way up to 50 kHz. Simultaneously, it will also produce a triangle wave. This type of wave form shows up distortion even to someone not overly familiar with the display. The lack of symmetry on a triangle wave cannot be ignored.

Circuit Description

The IC used is an LM3900, which consists of four separate, internally compensated amplifiers, which are also short-circuit proof. However, only two of these sections are used here. The first operates as an op-amp and the second as a Schmitt trigger. The output of pin 4 goes to the inverting input of the second section, which is pin 6. It also goes directly to jack J1, providing a triangle wave. The square wave is at jack J3, the output of pin 5. J2 and J4 are grounded to the chassis, while J1 and J3 must be insulated. This arrangement permits the two waveforms to be used at the same time. This renders a switch unnecessary.

Fifteen volts is the recommended operating voltage. A single-ended supply can be obtained in two ways: since only 10 mA of current is drawn by the unit, batteries can be used with a long operational life. You can also build a simple supply to fit within the unit. In this case, you would need a larger box to house the additional components. If you do not anticipate using the generator very often, you can provide jacks that permit you to accept the operating voltage from your bench voltage supply.

Output impedance for either square or triangle wave is 2 kHz. To simplify the construction of the unit, only one control was included...R5, which regulates the symmetry of the waveforms. For the same reason, no attenuation circuit was included. If you need some reduction of the output, you can use a probe with a 1-meg resistor in series.

APPENDIX 1
RESISTORS AND CAPACITORS

Resistors and capacitors are used in such numbers in electronic circuits that there is a tendency to take them for granted. The inexperienced builder may assume that all resistors are alike and any type may be used whenever that squiggly symbol appears in a schematic. While perhaps this is true most of the time, there are cases when the wrong type of resistor can be less than helpful and even create unsuspected problems. However, resistors are less apt to cause such confusion than capacitors—which come in a variety of types, shapes and functions.

For example: an electrolytic is indicated in the parts list and its plus and minus leads are shown in the schematic. So, that takes care of that.

But does it?

Slightly different capacitances can be substituted without much harm. Capacitors are high tolerance units, usually, and a 20 μF cap can be used instead of a 22 μF. What about the voltage rating? Let us say that the original circuit contained a 35-volt unit, or the schematic called for that working voltage. If the hobbyist throws in a 50-volt capacitor, will harm be done? No, not even if a 75-volt unit were put in instead. However, (and this is the point often ignored) the higher rated unit will generally be larger and therefore take up more room. This can be a problem when replacing a capacitor on a PC board, especially when a board has been designed with a specific size in mind. This is also why it is recommended that you buy all of the parts you are going to use before you begin to draw the PC layout.

So, a larger capacitance can be substituted for a smaller. However, it is a good idea to know first what the function is of that capacitor in the circuit.

Take for example, a capacitor used for filtering in a power supply, as in Fig. A1-1. A 2000μF capacitor in this case will give you better filtering than the 200 μF called for. Why was the smaller one specified? More than likely because this was what the author had on hand and it worked adequately. So in this case you can use the larger one.

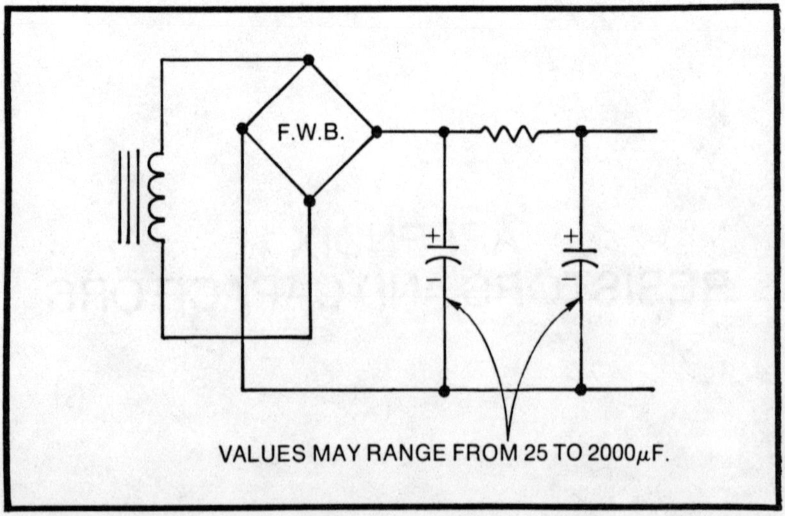

Fig. A1-1. Application of a capacitor in a filtering function. Example shows its location in a power supply.

What about a bypass capacitor? Figure A1-2 shows its use in a circuit. Its function is to provide a low impedance path for the AC current to ground. Again, a higher value can be employed. Still, it is better in this case to remain closely to the specified value. The designer of that circuit may have had a real reason for that value.

Still a third use for a capacitor is to join two circuits or two stages of the same circuit. (See Fig. A1-3.) The capacitor must also block the flow of DC, while allowing the signal or AC current to pass. When it serves such a purpose,

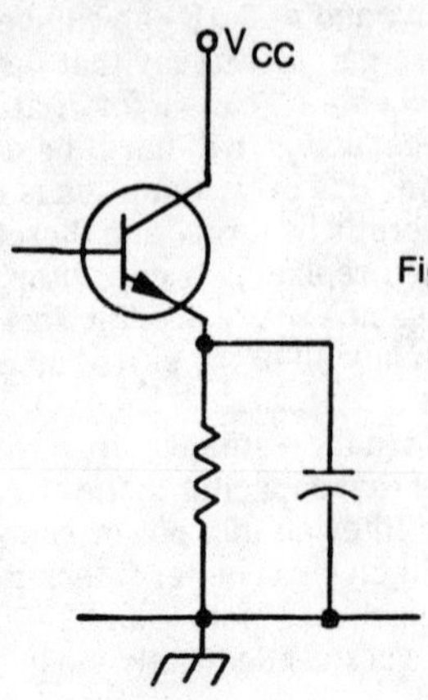

Fig. A1-2. A capacitor used for by-passing.

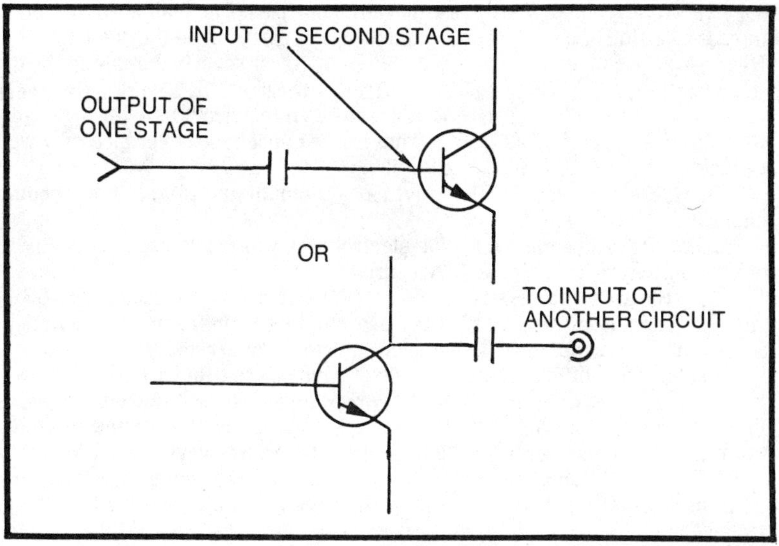

Fig. A1-3. Capacitors being used in coupling two parts of a circuit, or in blocking DC in an AC circuit.

it is called a coupling or blocking capacitor and there the value specified should be strictly observed. The reason is that the capacitance of the unit will have an definite effect on the frequency being passed.

TYPES AND THEIR USES

Typical questions that are often posed are: Why is a certain type of capacitor mentioned in a circuit? Why mica, ceramic, paper, or plastic film? What does each type do?

Ceramic capacitors are made up of a ceramic disc coated on both sides with metal. There are two types within this group: one is a high-dielectric ceramic, which offers high capacitance in a small size. Tolerance, which is stamped on the disc, or colored coded, will be from − 20% to + 100%. These capacitors are subject to value changes due to temperature variations, frequency, and applied DC, so they are used for bypassing where an exact value is not required. The second type, low-loss, has large leakage and is used mostly in high-frequency circuits.

Silver mica capacitors are stable with a low temperature coefficient and able to withstand electric stress. They have low leakage current and are available from about 1 pF to 0.1 μF with 1% to 2% tolerances. Their working voltage is usually 350 VDC. They are not polarized.

Paper capacitors range from 400 pF to 50 μF and can withstand high voltage. Apart from poor tolerance, they suffer from high leakage current. They also have a tendency to dry out with age, so the following type is often substituted for the paper.

Plastic-film capacitors can be specified as Mylar, Teflon, or polyethylene depending on which dielectric is used. These capacitors work at very high temperatures with capacitance ranging from 500 pF to as high as 10 μF.

Electrolytic capacitors are generally polarized. One lead is connected to the positive side of a power supply and the other electrode to the negative side. Reversing the connection will result in destroying, or severly damaging the unit. Capacitance in this type will vary from 1 to 500,000 μF. Working voltages range from 2.5 to 600V. Tolerances are not symmetrical (as with a resistor which may be specified as ±10%. That is, the capacitance tolerance may well be indicated as − 20% + 30% or − 10% + 100%.

There are two types of electrolytics: aluminum and tantalum depending on the material used in the anode.

There are also some non-polar electrolytics which are used generally in cross-over networks in speaker enclosures.

All of the above are fixed capacitors. Although none were used in this book, variable capacitors also exist. They are employed for frequency control in some signal generators and in tuning circuits for a radio receiver.

What are the different types of resistors and how are they used?

Carbon composition is the most common type resistor. This type is used to limit the current flowing through a circuit. Resistance, designated in ohms, will range from one ohm up to 22 megohms; tolerances vary from ±5% up to ± 20%. Units which higher wattage ratings are used as non-inductive loads for amplifiers. Stability is poor while the resistance will change radically at radio frequencies.

Wirewound resistors are more stable than the carbon type and are available in tolerances as low as 0.01% with power ratings as high as 200 watts. They should not be used at radio frequencies since they are highly inductive.

Metal-film resistors are specified in some circuits as "low noise". Their presence is demanded in low-level audio amplifiers where any noise originating in the resistor would be amplified. Resistance may run as high as thousands of megohms.

Carbon-film resistors offer very low values of resistance—sometimes as small as fractions of an ohm. They have low tolerance.

Variable resistors are used in almost every circuit that calls for a control. Variable resistors with slider arms are generally high wattage and their function is to serve as voltage droppers. But those most used are:

Potentiometers with rotary action wiper to control current and signal. Pots, as they are called, range from 10 ohms all the way up to a megohm. There are two types in this group:

Linear pots are employed when a uniform rate of control must be imposed on a circuit. A mechanical travel of 10% will equal exactly 10% of the resistance. This type is most often employed in instrumentation and in tone and balance controls in audio circuits. They can easily be calibrated.

Audio pots are made so, that for a great proportion of travel of the wiper, there is little change in resistance. As the rotation continues the larger percent of the resistance is crowded in the upper part of the travel. This type is used as a volume control, as it conforms with our hearing curve. The decrease in volume is logarithmic in percent of travel. This is why these pots are also called by that name.

COLOR CODING

Color coding for fixed resistors is found in Table A1-1. The complexity of reading capacitor codes demands a section to itself, which follows.

Table A1-1. Resistor Color Code.

Color	First Significant Number	Second Significant Number	Multiplier
Black	–	0	1
Brown	1	1	10
Red	2	2	100
Orange	3	3	1000
Yellow	4	4	10,000
Green	5	5	100,000
Blue	6	6	1,000,000
Violet	7	7	
Grey	8	8	
White	9	9	
Gold			divide by 100
Silver			divide by 10

No 4th band ±20%
Silver 4th band ±10%
Gold 4th band ±5%

The color coding for resistors is standardized and, since there is only one code to identify their values, before long the code is memorized. With capacitors however, it is not so easy. The complexity has two causes: there is no agreement among the manufacturers as to a standard code; there is also the fact that we need much more information on what a capacitor can do. All of these factors must be stamped in some way on the body of the component.

The capacitance can be marked in a code which is identical to that of the resistors—but only up to a point. The user also wants to know the tolerance and the working voltage. This is only the beginning. A manufacturer of a piece of electronic equipment also has an interest in the temperature coefficient, type of dielectric, and the operating temperature range.

As though the problem were not complex enough, it is further complicated by the many shapes and sizes in which capacitors come. They may be round, tubular, disc-shaped, miniature, and subminiature. Some are so tiny there is no room for markings. The container in which they're packed has the information; this is of little value once the capacitor is out of its case.

Paper, mica, and ceramic caps use a color code similar to that of the resistors. But, it should be remembered that the capacitance is marked in picofarads (pF = 0.001 μF). A mica capacitor is generally coded in the six or nine dot system. Much of the information carried by these dots is of little value to the hobbyist. He is not working to military specifications and his working temperatures are not apt to be significant, and will contribute little to the operation of the unit. So, a complete description of what each dot means will not be explained here, since it would take a whole section of the book to cover it

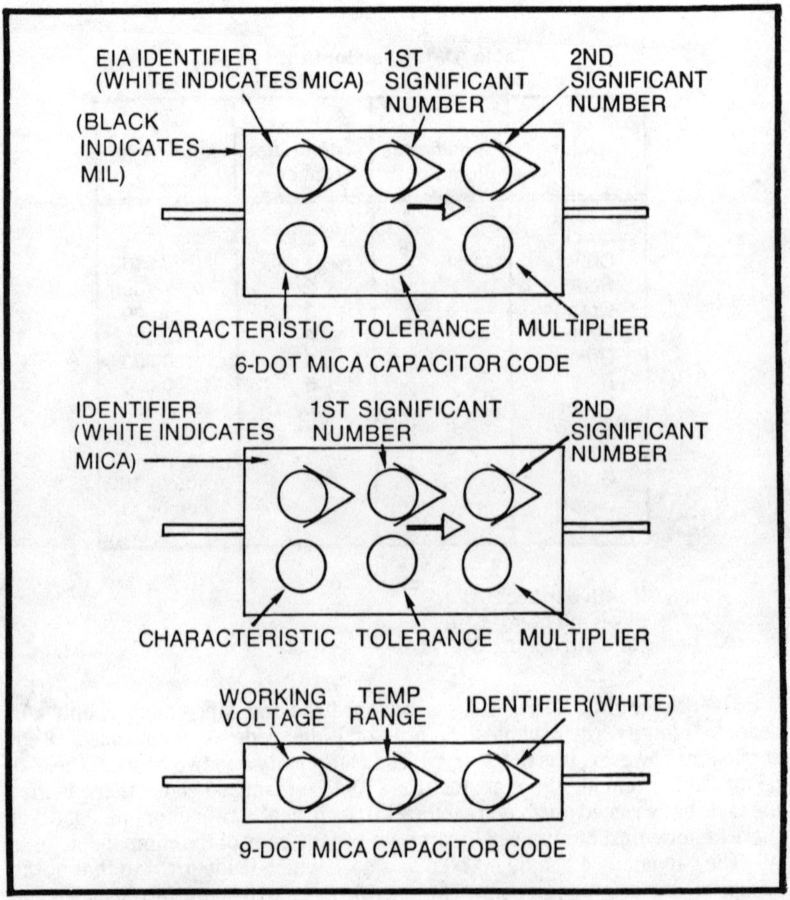

Fig. A1-4. Color code for resistors. Also simplified interpretation of a 6 and 9-dot color code for a mica capacitor.

adequately. Discussion is limited to the capacitance and the tolerance. With these two bits of information, the experimenter has enough to go on.

Some manufacturers stamp the value in numbers (not in code) of the capacitance and the tolerance on one side of the body of the cap and the working voltage on the other side. This is simple and neat and presents no problems for the user.

Unfortunately, many other manufacturers use the six dot system and here is how it works (see Fig. A1-4).

The nine-dot code offers basically the same information as the previous six-dot code, but both sides of the capacitor are used. The three dots on the back indicate working voltage and the operating temperature range, with the last dot as an identifier which repeats the information presented by the first dot on the front.

If you think you are confused by all of this. remember that this is only a fraction of the various sytems. What you must be able to do is have the ability to read the capacitance. This will give you an idea whether it is worth putting that capacitor on your capacitance meter to get the exact value.

The meter is of especial value when you buy surplus capacitors and they are marked in the MIL code. or the manufacturer uses a part number for his capacitor.

It is a good idea that. as soon as you buy a capacitor which uses one of these codes. and before you forget the value. mark the capacitance and tolerance as well as the working voltage and what the cap is on a piece of gummed tape. which is then adhered to the cap in question. When you are about to use the cap. it is simple to strip off the tape.

APPENDIX 2 INTEGRATED CIRCUIT PIN CONNECTIONS

This appendix indicates the pin arrangements for the ICs used in this book. In every case, we are looking at the IC *from the top*. This means that the numbers for the pins are running *counterclockwise*. In many magazines, the IC is viewed from underneath. In this method the numbers run clockwise. You will see that many of the ICs come in more than one type package. Some are round with the pins arranged in a circle and have a tag projecting from the body which tells you the starting point of the counting. Again, this may be in one direction or the other, so you must know which way that particular IC goes. The same IC may be also packaged in a rectangular package with the pins arranged in two parallel rows. A dot or a notch at one end indicates the starting point. But, due to the lack of standardization on the part of the manufacturers, the dot can be the high or the low end. So, you have to make certain by getting the specification sheet when you buy the IC. At least, you should obtain the pin number arrangement.

To correspond with their own numbering systems, manufacturers assign their own part number to an IC made by another with his own number. In most cases, the substitutes indicated in the following section will work just as well as the original.

Many of the ICs made for hobbyists come in "blister packs". Keep the sheet of paper or cardboard that comes with the component. It will have the pin numbering and other pertinent information.

Sometimes you will see that an IC—as do some transistors—will carry the part number followed by a letter; such as μA741A or μA741E. This means that the unit will meet tighter specifications for industry or the military than the run of the mill lot without the suffix letter. This information will be of little value to the hobbyist so he can use any of the components without hesitation.

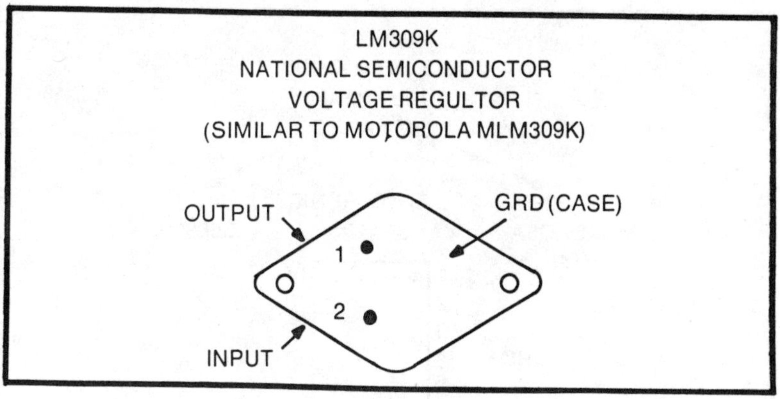

Fig. A2-1.

IC PIN CONNECTION DIAGRAMS

Figures A2-1 through A2-15 show common IC connections. All ICs are viewed from the top. Only those substitutes available at time book was written are listed (ICs are listed in numerical order—not alpha numerical).

SPECIAL SUBSTITUTIONS

Motorola has the HEP line which is made up of functional substitutes for many ICs, transistors, and diodes. They, too, carry a different part number. RCA and GE also have their own experimenters' or servicemen's substitution list. Their main advantage is that these lines are carried by most electronic supply stores. They have a tendency to be a little expensive, but since the average hobbyist buys only one or two units at a time, the extra expense is not

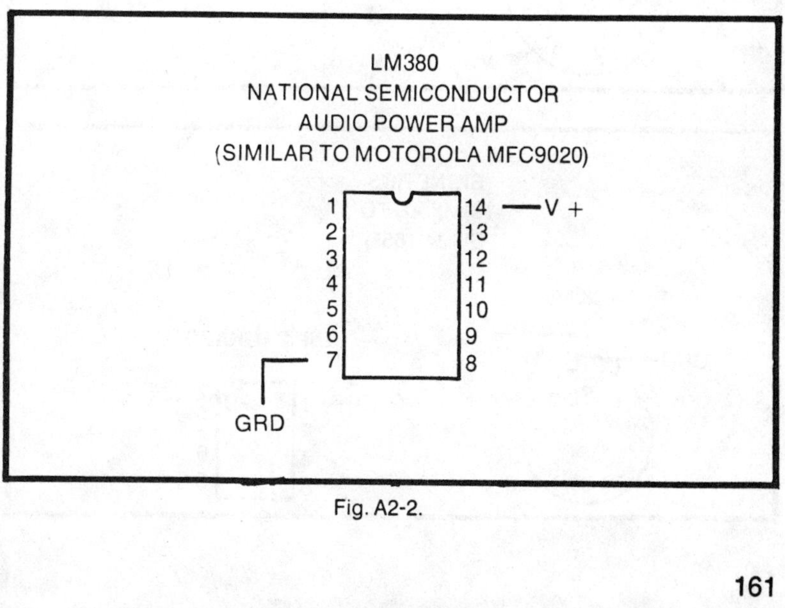

Fig. A2-2.

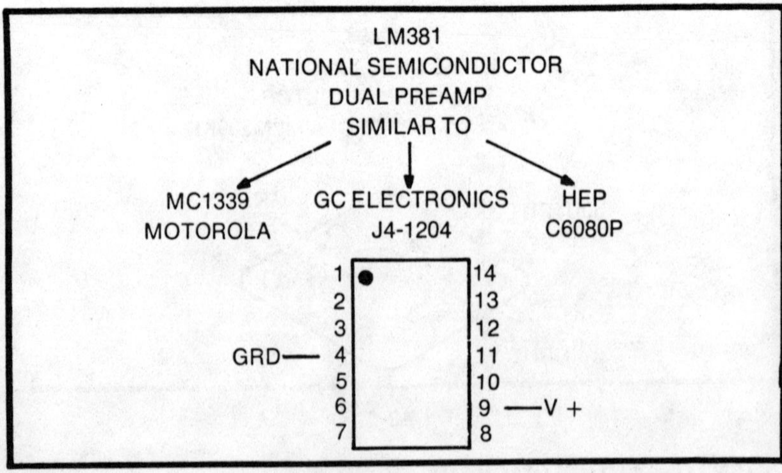

Fig. A2-3.

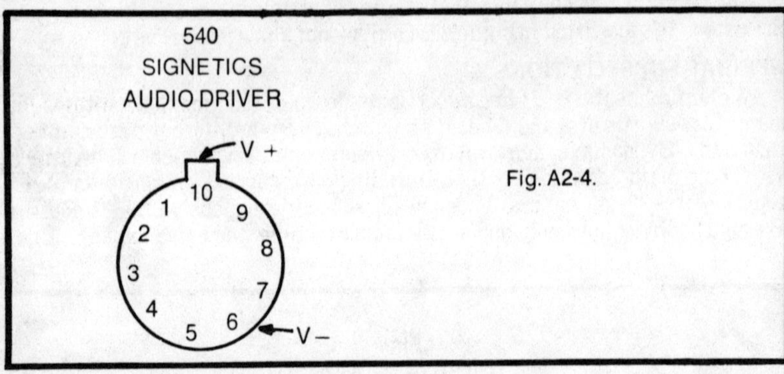

Fig. A2-4.

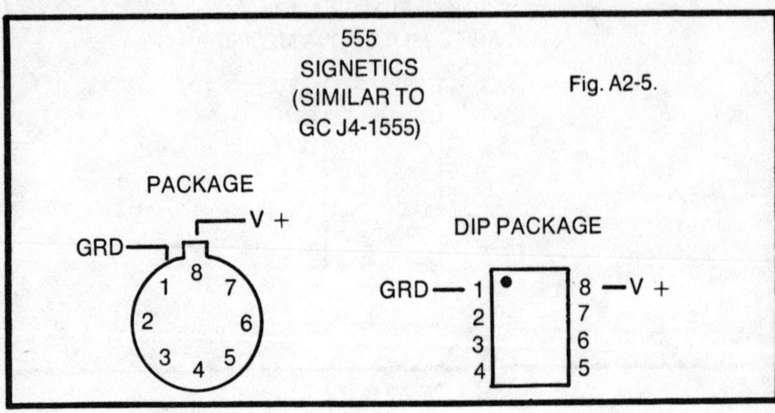

Fig. A2-5.

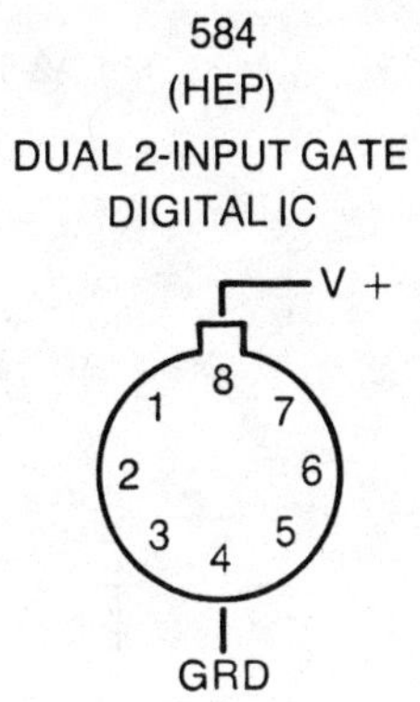

Fig. A2-6.

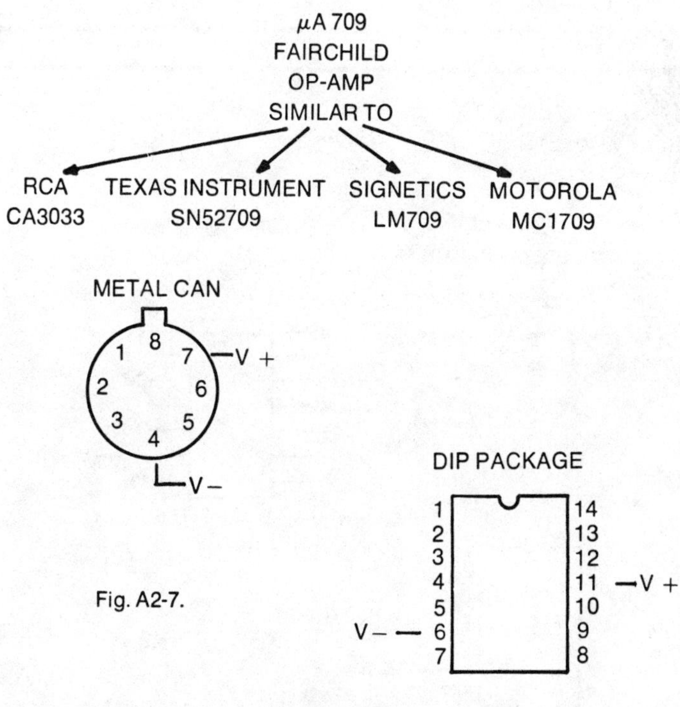

Fig. A2-7.

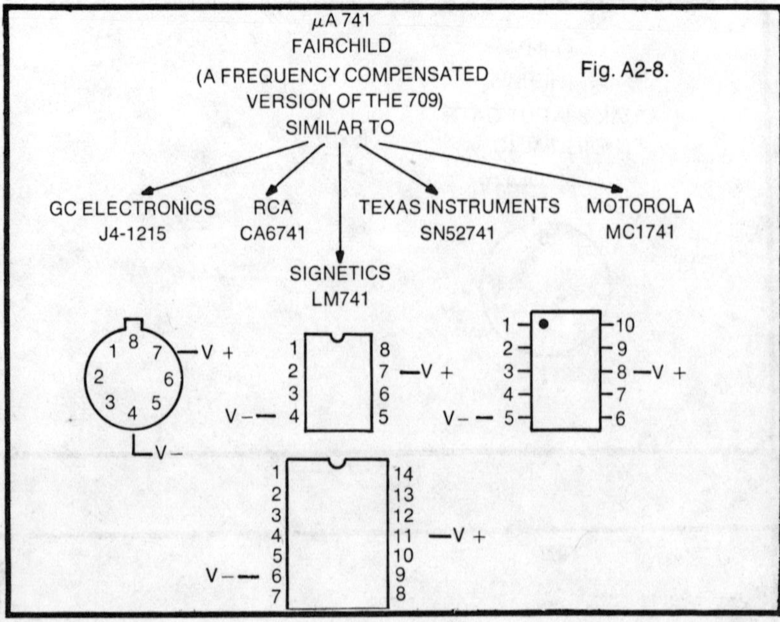

Fig. A2-8.

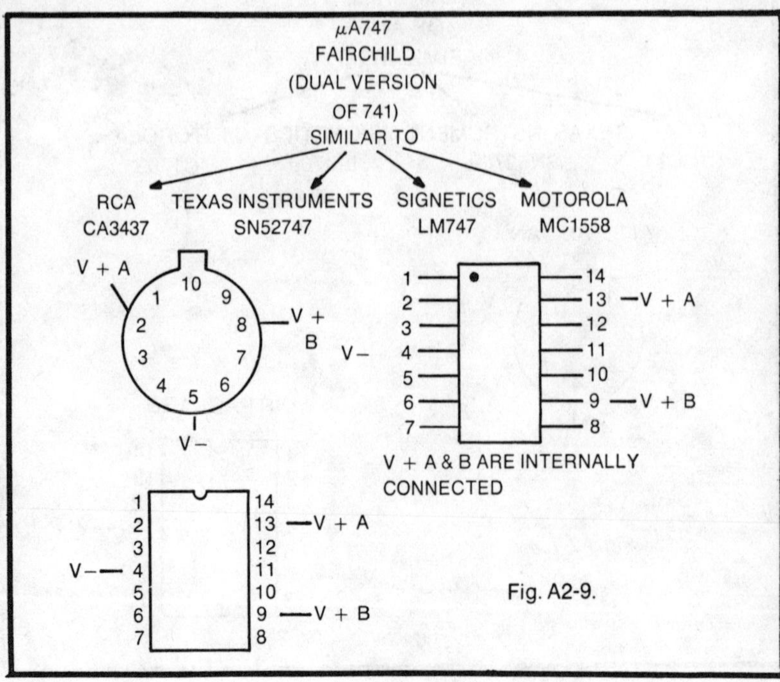

Fig. A2-9.

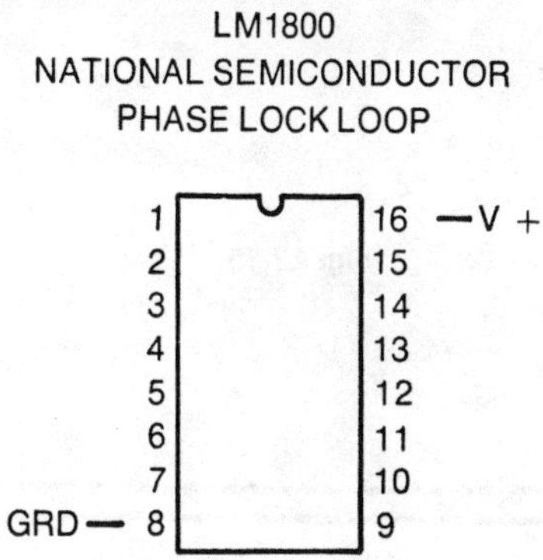

Fig. A2-10.

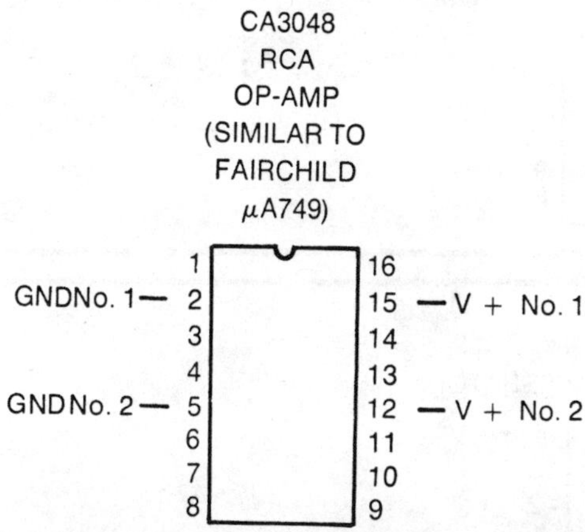

Fig. A2-11.

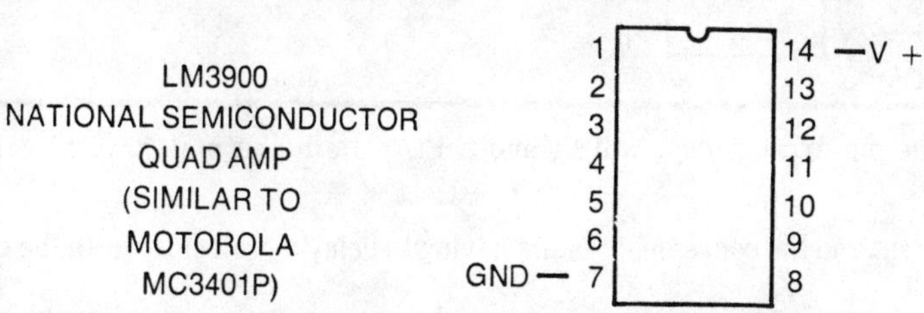

Fig. A2-12.

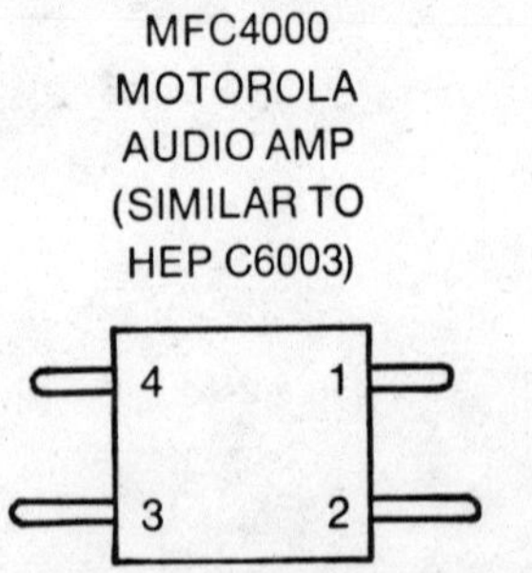

Fig. A2-13.

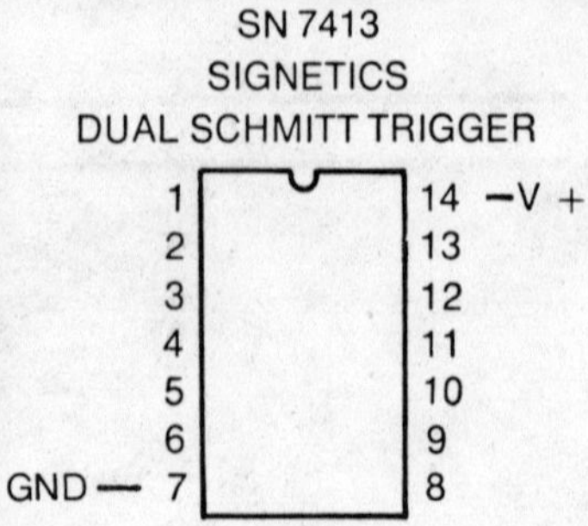

Fig. A2-14.

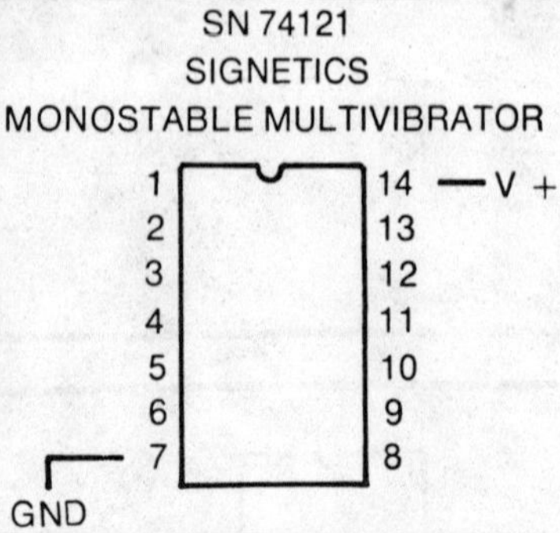

Fig. A2-15.

that the pin arrangement will fit into a PC pattern you may have already drawn!

that great and the convenience of not having to delay a project is worth the extra cost.

Their specifications may not be as tight as the specs of the originals. but except in the most critical areas it will make little difference. But. watch out

APPENDIX 3 MANUFACTURERS AND SUPPLIERS

The following list of mail order suppliers is meant as a service to the reader. It is by no means complete. These are the houses with whom we dealt in the past and whose products and services were satisfactory. This should not be misconstrued into believing that others not listed do not meet these qualifications. The list comprises houses offering a broad range of supplies. For that reason we omitted companies whose products were in one category only. such as a supplier offering only computer electronics or kits.

B & F Enterprises
119 Foster St.
Peabody. Mass. 01960

Cortland Electronics
114A W. Broadway
New York. N.Y. 10013

Edlie Electronics
2700 Hempstead Tpke
Levittown. N.Y. 11756

GC Electronics
Rockford. Ill. 61101
Their line is available
in many electronic stores

Lafayette Radio
111 Jericho Tpke
Syosset. N.Y. 11791

Olson Electronics
260 S. Forge St.
Akron. O. 44327

Poly Paks
P.O. Box 942
So. Lynnfield. Mass. 01940

Radio Shack
2617 W. 7th St.
Fort Worth. Texas 76107
They have thousands of stores in the U.S. and there is bound to be one near you.

RAMCO Electronic City
365 Canal St.
New York. N.Y. 10013

Solid State Sales
P.O. Box 74
Somerville. Mass. 02143

Most. if not all. of these suppliers will furnish you with their catalog on your request.

INDEX